MUSIC AND YOUR EMOTIONS

Music and
Your Emotions

A Practical Guide to Music Selections Associated
With Desired Emotional Responses

Prepared for the Music Research Foundation,
Incorporated, by

ALEXANDER CAPURSO, Ph.D.
Director of the School of Music
Syracuse University

VINCENT R. FISICHELLI, Ph.D.
Department of Psychology
Hunter College

LEONARD GILMAN, M.D.
Diplomate, American Board of
Neurology and Psychiatry

EMIL A. GUTHEIL, M.D.
Diplomate, American Board of
Neurology and Psychiatry

JAY T. WRIGHT, Ph.D.
Director of the Workshop in
Intergroup Education, Columbia
University Teachers College

FRANCES PAPERTE
Founder, Music Research
Foundation, Incorporated

LIVERIGHT

NEW YORK

SBN   87140-032-4
Library of Congress Catalog Card Number 75-131283
MANUFACTURED IN THE UNITED STATES OF AMERICA

This book is dedicated to the many sponsors of the Music Research Foundation without whose financial assistance the work of the Foundation could not be carried on.

The authors of this book have agreed that any and all profits accruing from the sale of this book shall be offered to the Music Research Foundation for the purpose of furthering research in this field.

ANALYTICAL TABLE OF CONTENTS

ANALYTICAL TABLE OF CONTENTS

INTRODUCTION

by Emil A. Gutheil, M.D.*

> "If we begin with certainties, we shall end
> in doubts; but if we begin in doubts, and are
> patient in them, we shall end in certainties."
> —*Bacon.*

The influence of music upon human emotions has been studied empirically for a long time. A large number of ingenious observations have been collected and recorded, in schools, clinics, hospitals, and industrial institutions. Experiments proceeded along two main lines: investigation of the effect music has on the physiological functions of the human organism, and study of its influence upon the human mind.

On the basis of these observations it was assumed that music can elicit certain psychosomatic reactions within the spheres of circulation, respiration, and others, and that it also can influence the mood and other psychological conditions in a large group of people. The investigators felt that these results warranted the expectation that music, properly applied,

* Chairman of the Research Committee, Music Research Foundation, Inc.; Director of Education, Institute for Research in Psychotherapy; President, Association for the Advancement of Psychotherapy; Editor American Journal of Psychotherapy.

might one day play a definite role among the supportive techniques of psychotherapy.

The relative ease with which this medium can be applied for the benefit of large groups of patients stirred up a great deal of hope and enthusiasm, not only in the ranks of musicians who expected from this application of music the opening of a new field for their professional endeavors, but also in medical circles and among the public at large.

The enthusiasm was somewhat cooled, however, when a number of pertinent questions were raised regarding the *principles* that were to govern the practical application of music, and the selection of the *musical and human material* for the individual therapeutic experiment.

"We want to apply music," said the enthusiast.

"Music?" asked the critic. "What music? Do you mean the melody, the rhythm, the harmony, the timbre, the overtones, the volume, the form, the vocal, the instrumental, the solo, or the ensemble? Do you mean music as it is practised or listened to? Or are you referring to a specific type of music, such as dance or march music, popular or classic, a lullaby of an Indian squaw or a symphonic poem by a Finnish composer? Are you talking about a full composition or a fragment of it, and if the latter, how much?"

"We want to study the *effects* music can have on *human beings*," the enthusiast responded in a considerably lowered voice.

"Effects?" sneered the critic. "Do you refer to the

physical, the emotional, the behavioral, social, educational, imaginational, quantitative, qualitative, integrative, synesthetic? Which of these effects are you going to study, measure, classify—and how are you going to do that? And as to human beings—what *sort* of human beings do you have in mind? What is their sex, personality, nationality, background, constitution, training, musical tradition, state of health, educational level?"

If we realize that each of the queries mentioned represents at least one of the unknowns and variables in any plan for a scientific evaluation of the effects of music, we may wonder if our research can ever reach the goal of science, namely, to *discover laws of cause and effect* in order to *predict the results*.

There is no doubt, however, that the job can be done. Carefully planned and properly controlled experiments must be conducted at various levels, involving the two main fields of exploration, that of music and that of the human mind. Many musically trained psychiatrists and psychiatrically-minded musicologists are at present engaged in this investigation. Some of their work deals with more basic problems, such as how music and the human mind communicate, while other work concerns itself with more specific inquiries, such as the relation of music to analgesia, anesthesia, fatigue, muscle tone, pulse rate, and respiration. The subjects include normal and abnormal individuals, adults and children.

It is as yet too early to survey the results; but one conclusion stands out: namely, that any scientific research directed toward the exploration of the above questions must (1) adopt a long-range plan of investigation; (2) break up the entire research into a large number of part-projects; (3) establish strict and uncompromising controls for all detail projects; (4) establish a network of research groups spread over many parts of the nation and beyond its frontiers, to correlate and consolidate the scientifically validated findings into a useful mosaic of knowledge that can serve as a substratum for a future therapeutic use of music. This entire plan is based upon the conviction that *one* carefully observed and recorded clinical fact weighs more than volumes of glib speculation on the value of "music therapy."

This book attempts to do three things: (1) to introduce the reader to the broad problem of the relationship of music to emotions; (2) to outline the scientist's quest for a suitable approach to the problem; and (3) to present the results of two major studies designed to relate mood changes to particular musical stimuli. The pioneer studies of Frances Paperte and her associates at Walter Reed General Hospital offer convincing evidence of the close tie between music and emotional states in a patient population. The Capurso studies growing out of this research have now extended the area of investigation to the so-called normal population and, in addition, provide us with empirically verified listings of specific musical compositions and the moods they are likely to mediate in the listener.

With all due reservation, it can be stated that the data collected here have a great practical value for workers in the field and that the Capurso listings constitute a necessary first step toward establishing an index of musical selections of known psychological value. The reader is invited to join us in our great experiment by testing further the validity of Dr. Capurso's findings.

We are still working under the handicap of over-anxiousness on the part of large groups of the public who expect from science a prompt delivery of perfect therapeutic implements. We are, therefore, appealing to the impatient to bear with us. The magnitude of the problem and the intricacy of the minutely detailed investigations place upon those concerned with research a burden of responsibility that can be allayed only through intelligent co-operation with and cheerful support of the many big and little projects by an enlightened, sympathetic, and discriminating public.

The concluding section of this book is designed to acquaint the reader with the Music Research Foundation, under whose sponsorship the studies reported on here were completed. A summary description of some of the more important research projects currently in progress under grants by the Music Research Foundation is also given.

THE SCIENTIST'S QUEST

by Vincent R. Fisichelli, Ph.D.* and Frances Paperte**

> With things investigated knowledge be-
> comes power.—*Confucius.*

A writer with a layman's point of view might find it a relatively simple matter to define with confidence the psychological phenomenon called "emotion," particularly as it applies to the way in which one reacts to music. He might find it completely adequate to attempt to describe a variety of emotional re-actions to music by the use of a series of phrases such as "Stravinsky disturbs," "Wagner exalts," "Sibelius depresses," "Chopin stimulates," and so on. Surely these statements would find agreement in some quarters, and because of that fact the writer could revel in the realization that, according to one of Poe's criteria of a good author, he had aroused in his reader the feeling that a common experience had been shared.

It is obvious to the reader, however, that it might be equally accurate to maintain that Stravinsky stimulates rather than disturbs, or that Sibelius exalts and Wagner depresses. There is little doubt, further-

* Department of Psychology, Hunter College.
** Founder of Music Research Foundation.

14

more, that these latter statements would also find their share of adherents. It is precisely in this apparent quandary that we find the crux of the matter: The emotional reactions which music may arouse are as numerous as the individuals reacting, and the subjectivity of emotional experiences which is reflected in all this variability is the core of our problem. Unfortunately, however, if subjectivity is the major problem with which we have to deal in order to understand the relationship of emotional experiences to music, it is at the same time the dominant obstacle in the way of arriving at that understanding from a scientific point of view. Science is usually befuddled by subjectivity. Its object is communality, not individuality. It looks for common causes, common relationships, and common effects. It seeks to uncover general principles, and all its tools are geared to the ultimate production of statements regarding the probability that a particular event will occur under certain conditions because of its frequent occurrences under those same conditions in the past. It is thus assumed that all of the repeated occurrences in the past are identical, or else the word "repeated" has no meaning; and the consummate distrust of unique events is reflected in the dictum, *scientia non est individuorum*.

But what could be more unique or more individual than what one person says he experiences? And this is what is meant by subjectivity. This is the same ogre the scientist likes to avoid and frequently refers to as the *"error* of subjectivity." The reference, of course, is not without justification. It does not require much study to see the possible error that may be committed in the complete acceptance of one person's experiences as typical of other persons' experiences. The famous story of the astronomer Maskelyne presents an interesting indictment against subjectivity and is perhaps worth repeating here, since it illustrates the manner in which science has come to recognize one of its limitations.

It appears that in 1796, Maskelyne, working at the Greenwich Observatory, was engaged in measuring the speed of transit of certain stars. His observations were of the greatest importance, since they were to be used in the calibration of the clock and subsequently in all other observations of time. His method was the then widely accepted method of Bradley, which involved counting the seconds spent by the star in traversing a known distance. The seconds could be counted with the beats of a clock or a metronome. The known distance was actually a space in the field of the telescope which was divided by parallel hair-lines; thus, the counting was begun when the star was seen to cross the first hair-line and continued until the star crossed the second line. If a beat was not exactly coincidental with the star's crossing of a line, estimations in tenths of a second were made. As was customary, Maskelyne employed a competent assistant, whose job was to make independent observations of the same transits. The assistant, named Kinnebrook, found, to his chagrin, however, that his recordings were at variance with those of his master's. The discrepancy amounted to some eight-tenths of a second, and it soon became apparent that someone was committing a constant error. Kinne-brook was dismissed.

Unfortunately for Kinnebrook, the event of his dismissal went unnoticed for some seventeen years, when finally another astronomer at Königsberg, named Bessel, read of it in a publication on the history of the Greenwich Observatory and began a systematic investigation of these personal differences. His studies are considered today to be pioneer investigations in the Psychology of Individual Differences. Briefly, his work led to realization that individuals differ in the way they experience the world around them, and that they have no control over these discrepancies, which are basically involuntary.

The moral of the above story is simple: If we cannot accept with certitude any single person's observation of a relatively simple and objective phenomenon because he is likely to be at variance with some other observer, what shall we make of the tremendous variability which meets us when we examine many persons' observations of the way in which they react to music, or even to a simple tone?

Compare the task of the astronomers described above with, for example, the estimation of a difference between two notes struck on a piano keyboard.

Time, broadly speaking, has only one dimension: duration. Tones, on the other hand, are multidimensional and can be characterized by pitch, loudness, and duration, to say nothing of the quality peculiar to the tone-generator which has been called timbre. Other dimensions that also have been suggested are brightness, vowel quality, volume, and density. The estimation of time requires a *quantitative* judgment: the problem is not what *kind* but how much. The matter of what is involved in the estimation of tone is controversial to this day and is complicated especially by the factor of interdependence among the psychological dimensions and the physical characteristics of tone. Thus, while loudness increases with the intensity of the tone, it varies in a complex fashion with the frequency of the tone; and while volume increases with intensity, it decreases with frequency. Finally, time estimations have reference points that are based upon widely accepted and familiar units such as the day, the clock hour, the minute, and the second. This is not true of *heard* differences in tones, except perhaps for expert musicians of long experience who may have constructed their own reference points for heard tonal differences. Surely, it appears from all this that the second task is incalculably more difficult than the first. Actually it has not yet been determined whether, from a psychological viewpoint, tonal estimations are more difficult to make than temporal ones. What *is* obvious is the fact that we ask of the

18

astronomers a *quantitative* question (How much?), in one dimension (duration), in an area where everyone has had much experience (time); whereas in the second case we ask a *qualitative* question (What difference?), concerning something which has many dimensions (loudness, pitch, etc.), and regarding which there is variable familiarity (musical tone).

We could easily reverse the tables so that the scientist could measure with ease the tonal problem and suffer with the temporal one. Let the first problem be: If the pitch of two tones and their duration and timbre are held constant but their intensities differ, which is louder? Let the second problem be: Does an hour in the office pass as quickly as an hour at the club? Thus it is evident that the problems of subjectivity which the scientist meets even in simple observations are quickly multiplied when he must analyze qualitative aspects of a multi-dimensional problem in persons who have little experience or training. The problem of how people react to music becomes infinitely complex in this light.

The cause is not entirely lost, however. Science has grappled with the problem for a good many years now, and because of this apparently overwhelming difficulty, rather than in spite of it, the issue is no longer still in doubt. The studies reported in this book are convincing evidence that a way can be paved successfully if we (1) are aware of our limitations, (2) set down our definitions carefully, and (3) restrict our conclusions and generalizations to the singular conditions of any particular investigation.

The mere listing of these procedural rules, however, does not solve problems for the scientist. He must convert them into efficient operational devices that ultimately can produce verifiable observations; and where an understanding of the problem of feelings, evaluation, and emotions is concerned, he has already met with some success. Let us examine the course he has charted for us.

In the first place, he has taken into account the fact that a person's testimony, however subjective, can be an important source of information regarding emotional experience, and accordingly he has carefully developed a series of techniques designed to get that information with a minimum of error. These methods are called the *methods of impression* and are concerned with the conscious feelings that a particular stimulus arouses. They involve a variety of specific techniques, each of which has its own area of application and is calculated to overcome some particular obstacle. Thus, where judgments regarding a complete musical selection are required, as in the Capurso study reported in this book, the *method of single exposure* requires that the subject listen to one piece at a time and then judge it by means of a scale offered to him by the experimenter. In the *method of paired comparisons* the subject is given two stimulus objects at a time and is required to indicate his choice. Obviously, where such materials as two musical chords or two tones are offered, the presentations must be successive. The *method of choice* involves the selection of one stimulus from many that are presented, and the *method of order of merit* demands that the subject rank in order of preference the variety of stimuli presented to him.

In addition to the *methods of impression,* science now has available an ever growing number of precision instruments for the objective study of emotions by the so-called *methods of expression.* These are directed at measuring bodily changes in the individual while he is emoting, and included among them are a large number of instruments borrowed from physiologists. There are devices called psychogalvanometers, for measuring changes in electrical potentials on the surface of the skin. There are measures of changes in respiration rate called pneumographs, measures of tremor, blood pressure, body sway, muscle fatigue, pulse beat, involuntary movement, and so on.

In the beginning, the aim of studies on bodily changes during emotion was the discovery of specific changes that might be said to accompany particular emotional states. The attempted tie-up between specific changes and specific emotions has failed, but there remains the possibility that more refined techniques and more precise definitions will one day help us to uncover such a relationship.

Finally, the scientific investigator of emotions has engaged himself in a continuous and systematic search for *variables,* or factors related to his problem.

Indeed, in this respect, he has made tremendous strides; and the actual research which usually grows out of such an enumeration of related factors lags far behind him. The Appendix to this book contains an example of what one investigator considers to be the vital factors related to the study of music and its effects on behavior. The investigator, Dr. Gutheil, has carefully listed in a systematic way the specific areas in which more information is required before even the vaguest spark of light can be thrown on the entire problem of the effects of music on behavior. The reader will observe that he touches on four major dimensions of the question: Materials, Effects, Listeners, and Objectives. The compendium, which has been conveniently called MELO, should provide the working basis for organized research in the field, in spite of the fact that Dr. Gutheil believes it to be still in a tentative and incomplete form.

A concluding observation which forces itself upon any serious researcher in this field is the fact that the advanced research is being guided more and more by *statistical* controls. The studies by Dr. R. Cattell now under way, described briefly later in this book, provide an excellent illustration of the way in which the latest statistical devices may be put to work in contributing toward an understanding of the problem of personality and music.

In brief, then, it might be said that our task is evident. The problems have been delineated with sufficient clarity, and the experimental methods have been tried and tested. What remains is the active and persistent implementation of research to the end that information on the relationship between music and man will provide a new key to understanding human behavior.

> ... music is a higher revelation than all ... wisdom and philosophy.
>
> —*Beethoven*

MUSIC AS A PSYCHOTHERAPEUTIC AGENT *

*by Leonard Gilman, M.D., and Frances Paperte ***

> The man that hath no music in himself
> Nor is not moved with concord of sweet
>   sounds,
> Is fit for treasons, stratagems, and spoils;
> The motions of his spirit are dull as night,
> And his affections dark as Erebus;
> Let no such man be trusted. Mark the music.
> —*Merchant of Venice,* Act V, Scene I

I. INTRODUCTION

Through the ages the influence of music over mind, body, and emotions has manifested itself so diversely and so frequently that its existence is no longer denied. In response to harmonious sounds, recorded changes of moods, influence on appetite, sleep, and general well-being have occurred too often to be ignored or taken for granted. A second devastating world war, which left our neuropsychiatric wards

* Reprinted by permission from The Journal of Clinical Psychopathology, Volume 10, No. 3, July, 1949. Copyright, 1949, by the Washington Institute of Medicine.

** Leonard Gilman, M.D., was formerly chief of the Psychiatric Section, Walter Reed General Hospital, and is a Diplomate of the American Board of Neurology and Psychiatry.

Frances Paperte is the founder of Music Research Foundation, Inc., was for 3½ years director of the Department of Applied Music, Walter Reed General Hospital, and is a former prima donna of the Chicago Opera Company.

filled with patients pathetically out of gear with reality, spotlighted the urgent need for discovering and developing new methods of controlling human emotions and behavior. To the end that music's potentialities be authoritatively investigated, Music Research Foundation was organized, and its initial studies were begun in 1944 at Walter Reed General Hospital, Washington, D. C., upon authorization of the Surgeon General's Office.

This report is based on three and one-half consecutive years of constant study of the use and effects of specific music in relation to mental patients at the Army Medical Center, Washington, D. C. It is hoped that recorded observations with accompanying medical evaluations on the use of music with patients representing a cross-section of mental disease, will serve to orient others interested in further scientific investigation of a vast new field.

II. HISTORICAL BACKGROUND

A. *Commentaries from Philosophy.* There is much in recorded history to justify an investigation of the influence of music upon the human mind and particularly its possible uses as an adjunct to therapy. Music has been prescribed and used for various ailments for centuries. Literature glitters with "miraculous" instances of "healing with music." More often than not, unfortunately, these stories obscure the genuine therapeutic potentialities of music behind an aura of superstition, and the observations recorded are highly anecdotal. Far from throwing a helpful light on a beclouded subject, they are the basis for fabulous and mostly unfounded claims.

Much more convincing is the belief of the ancient philosophers in the efficacy of music over mind and body. For instance, Confucius [1] not only loved music but ascribed to it social virtues. He believed that ritual and music were the clues to harmonious living. In the Fourth Book of *The Republic,* Plato [2] states that health in body and mind which controls and improves the body is to be obtained through music and gymnastics and should continue throughout life. Aristotle [3] ascribed the beneficial and medicinal effects of music to an "emotional catharsis," a view subscribed to by many psychiatrists today.

Cassidorus [4] attributed to music the power to expel the greatest griefs, and said "it doth extenuate fears, furies, appeaseth cruelty, abateth heaviness, and to such as are watchful it causeth quiet rest; it takes away spleen and hatred—it cures all irksomeness and heaviness of soul." Pythagoras [5] gives us probably the most explicit explanation of the application of music as therapy in ancient times. His recorded experiences in the field furnished the inspiration for this approach applicable to modern man and geared to his needs. Pythagoras was of the opinion that music contributes greatly to health if used in an appropriate manner.

Coming down to us through the wisdom of the ages, this unbroken thread of belief in the efficacy of music is far more convincing than the "miracles" cited as music cures by modern investigators.

B. *Observations of Modern Investigators.* Music's functional potentialities, like those of electricity, are known to us chiefly through observation of the accomplished manifestations. Of its nature and its way of working, we know very little as yet. In spite of the intangible and elusive nature of the "x-factor" in music, scientists for centuries have been intrigued by its potential contribution to the body of applied science. Recent investigative studies on the sensorium have served to broaden rather than restrict the scope of music's effectiveness as an auditory stimulus. The intricate connections of the auditory pathways are still being unravelled.

The bulk of available information on the physiological effects attributed to music is contained in the following survey of investigations in the field. Many sporadic investigations seem to have been carried to a certain point and then abandoned. Among early attempts, Dogiel [6] was probably the first to experiment on human organisms by means of the plethysmograph (an instrument for determining and registering variations in size of an organ or limb, and hence in the amount of blood occurring there).

According to Binet [7] and Courtier, consonant and dissonant chords, major and minor intervals, changes in intensity, etc., all produce changes of pulse and respiration.

Other investigations confirming the effect of music

on pulse and blood pressure have been conducted by Gretry, Hyde, and Shalapino.[8] Weed,[9] experimenting further on organic phenomena present during musical attention, found that disturbances of the blood supply and irregularities in respiration both in rate and amplitude vary directly with the intensity of the experienced emotion.

Diserens,[10] in his investigation of metabolic changes, states that rhythm of respiration tends to adapt itself to the rhythm of the music, especially when the latter grows slower. The intensity of peripheral vaso-constriction depends upon individual impressionability.

Diserens' investigation also included tests of harmonious combinations of sounds upon the sensorium, and demonstrated that music is capable of lowering the threshold of sensory perception. Experimenting along the same lines, Kravkow [12] discovered that music and rhythmic sounds can improve a listener's eyesight as much as 25%. As little as the rhythmic ticking of a clock, experiments showed, served to stimulate the vision.

Experiments of Fere, Tartchanoff, Diserens, and Scripture [13] included studies of the effects of the sound stimulus upon the skeletal muscles. Using musical selections as the stimulus, Tartchanoff observed that (1) music exercises a powerful influence on muscular activity, which increases or diminishes according to the character of the melodies employed; (2) when music is sad or of a slow rhythm, and in the minor key, the capacity for muscular work decreases to the point of ceasing entirely if the muscle has been fatigued from previous work. The general conclusion is that sounds are dynamogenic or that muscular energy increases with the intensity and pitch of the sound stimuli. Isolated tones, scales, motifs, and simple tonal sequences have all been found to have an energizing effect upon the muscles.

Investigations made by Dutton and Tartchanoff [14] indicate that music increases metabolism and influences internal secretions. As to specific metabolic functions and their reactions to music, much experimental work has been done but much more specific work must be done under controlled conditions. Most physicians simply maintain that music exerts a strong influence on the higher cerebral centers and thence, through the sympathetic nervous system, upon the other portions of the body, thus promoting digestive, secretory, circulatory, nutritive, and respiratory functions. The favorable results are due to the agreeable occupation of the higher centers with music, so that the sympathetic nervous system is unhampered in its activity, while at other times the higher centers interfere with its supply of energy.

C. *Observations on the Relation of Music to Emotions.* Various theories have been advanced to explain the relationship between music and emotion. Hanslick [15] has said that "Music evokes feelings that are emancipated from worldly affairs. . . . Feelings not that of daily life . . . a general feeling-state or mood.

A person attains a state of consciousness free from worldly associations. A person is taken out of himself, . . . which is a wholesome experience."

According to Hanslick, Cannon, Dearborn,[16] and others, the emotions have been ascribed to images and associations aroused in the mind by music, the connection between emotion and music being indirect.

Richard de la Prade [17] noted that music is the only art to which animals, the feeble-minded, and idiots are sensitive, and that it, therefore, has in it an entirely physical element, a kind of electricity which affects the nerve independently of any action on intelligence. Davison [18] tells us that music undoubtedly exercises its influence over the body without influence on the highest nervous centers, and that the human organism participates in the tendency to vibrate synchronously with music, which is general in the animal world. Even deaf mutes respond to musical sounds. They distinguish several different musical instruments according to the nature of the vibrations.[19]

Science is well aware of the need for further information upon the affective value of sound stimuli. Diserens [20] says, "Psychologists and psychiatrists have a very real and pressing problem in the way of investigating and reporting on both the integrating and disintegrating effects of noise and musical stimuli." Schonaur [21] makes clear that an increasing volume of sound in modern life—without adequate control of its character—is one of the causes of growing emotional instability in contemporary society.

The objective data so urgently needed to throw light upon the value of music in psychology have been more difficult to accumulate than those pertaining to the physiological. Here, the research scientist must "grope" in the same sense as must the psychiatrist in his efforts to evaluate his chief tool, the spoken word.

Bingham,[22] in an experiment based on data obtained from 20,000 persons, reported the effects produced upon their moods by a variety of 290 phonograph records. The conclusion from these data was that a musical composition not only produces a mood change in the listener but it induces a markedly uniform mood in a great number of persons in a given audience. The feelings most frequently excited by music were those of rest, joy, sadness, love, longing, and reverence, while such negative emotions as anger, fear, jealousy, and envy were conspicuously absent.

Henlein[23] conducted experiments to determine the difference between the effects created by major and minor chords, and much study has been given to the effects of the elements of musical structure, namely, tempo, pitch, rhythm, harmony, and melody. Based on a study of this nature made by Kate Hevner, some conclusions were that in causing excitement, the most important element is tempo, and the tempo must be swift. Complex harmonies, lower pitch, and descending melodic patterns contribute much to this feeling of excitement. Still another factor is dotted figures or uneven rhythms. In producing a given affective state, melody plays a very small part in expressing musical meaning. The difference in expressiveness for certain melodies is generally attributed to their rhythm, tempo, etc., instead of the pitch pattern of the successive tones. Modality is very important in the dimensions of brightness, sadness, playfulness, and happiness, but useless in dimensions of dignity, vigor, excitement, and calm.

A similar study made by Gundlach [24] reported that tempo was by far the most important factor in arousing emotion in the listener, rhythm was second, and the melodic range was least significant.

Tonal and rhythmic patterns have been found to affect the human organism on all levels: on the hypothalamic level, evidenced by the instinctual response; on the cerebellar level, affecting coordination, equilibrium, and bodily rhythm; on the cortical level, stimulating imagery and association; and on the psychic level, evidenced by creative or aesthetic response. In patients who suffer from a state of cortical inactivity, and thus fail to respond to verbal stimuli, the cortex can still be invaded, it has been found, by using music to reach the patient at the thalamic level, after which the impulse is carried to the cortex apparently via the corticothalamic pathways. This fact is of considerable significance in dealing with otherwise inaccessible patients.

D. *Symbolism in Music*. In the vast bulk of psycho-analytic literature, relatively little can be found concerning the interpretation of the psychologically symbolic nature of music. Isolated efforts to interpret various phases of musical experience have been reported. Montani [25] in a brief, undocumented report divides the emotional responses in two main categories, the joyous and sad. He explains the "mysterious feeling of a sadness developed by the Minor Mode" as feelings which are associated in the unconscious with self-punishment or masochistic demands arising from the basic castration complex. Tilly,[26] in a more comprehensive report, discusses the "psychoanalytical approach to the masculine and feminine principles in music." She classifies the criteria by which she divides musical composers into a masculine group and a group possessing "neurotic feminine qualities." Evaluation of the works of 6 outstanding composers in the light of these criteria are compared to biographical data concerning the lives of these composers, and she concludes that the underlying psychological pattern of the man and of his work are similar. She illustrates this by citing Chopin, Tschaikowsky, and Liszt as "feminine," and Bach, Handel, and Beethoven as "masculine." She has observed that individuals respond most favorably to music (and hence to the composer) in which the degree of latent homosexuality (Anima) approximates most closely that of the listener. It is of course only to be expected that a man in whom the "Masculine Principle" is weak would show a preference for music which is strongly feminine in character, just as we would expect his unconscious homosexual component to dictate similar "feminine" tastes in clothing, colors, etc.

In 1922 Pfeifer [27] claimed that music provides a method of escaping reality through its basic rhythm, which preoccupies the consciousness to the degree that unconscious fantasies are released. He proposed that music is "pure libido symbolism lacking objectification" or cathexis.

Frequently in the course of psychoanalysis there is association of musical memories to events of the past and to elements in the present dream and fantasy life. Certain tunes become actual substitutes for fantasies associated with either pleasant or unpleasant thoughts, and these may at times serve as a key to understanding the repressed elements of an unconscious conflict.

III. SOME RECENT CLINICAL USES OF MUSIC IN PSYCHIATRY

In the treatment of mental illness, the following properties of music have been classified for their practical application: (a) the property of attracting attention and prolonging its span; (b) the property of producing various moods; (c) the property of planting engrams, stimulating associations, and imagery; (d) the property of relieving internal tension; and (e) the property of facilitating self-expression.

Similarly, it is generally agreed that there are three categories of responses a mentally unbalanced or emotionally maladjusted individual makes to musical stimuli: (a) through the rhythmic stimuli, muscular tensions are set up which seek an immediate release through physical motion, which may help to pull a patient out of morbid fancies or draw his attention to things happening around him; (b) the emotional response, which is the awakening of various moods in patients by different types of music; (c) the associative response, whereby music stimulates the process of thought, of both memory and fantasy formation, which makes it evident that such stimulation may facilitate the expression of repressed or unconscious mental elements.

Practical application of the aforementioned knowledge is in an embryonic stage. A survey of recent efforts to use music in psychiatric hospitals was reported in September, 1944, by Mrs. H. G. Price, in an unpublished communication. This revealed that music was being used in one form or another in mental institutions in at least 31 states as well as Canada and Hawaii. The more extensive programs at the time were being conducted at the Pennsylvania Hospital in Philadelphia; the Allentown State Hospital in Allentown, Pa.; the City Hospital in Cleveland, Ohio; Michigan State College; the Texas State College for Women; and the Ontario Hospital, London, Ontario.

A national survey of music in mental hospitals

made in 1944 by the National Music Council revealed that only 23 out of 187 institutions consider their own use of music to be therapeutic and not merely recreational.

Moissaye Boguslawski [30] has reported success in playing for patients in the Chicago State Hospital. His method depends chiefly upon the property of music to stimulate memory associations in the individual patients. He begins with nursery melodies, progressing through music of the primary, adolescent, and adult stages of an individual's life on the theory that a selection will thus be found which will penetrate the patient's world of fantasy and reestablish at least temporary contact, through memory association, with the outside world.

Dr. Bender,[31] psychiatrist at Bellevue Hospital in New York, has used music to correct personality faults in maladjusted children in the Children's Psychiatric Ward. "I am quite convinced," Dr. Bender has stated, "that our music activity reaches the subcortical centers of the brain, where other activities do not, and thereby helps to integrate the personality that is going to pieces in these children."

Music has also been used with mental patients at Eloise Hospital and Infirmary, in Eloise, Michigan, under the direction of Ira M. Altschuler.[32] Dr. Altschuler believes group singing around a piano in a ward is valuable in arousing and developing attention, as well as providing an opportunity for self-expression for as many individuals as possible. He considers that patients stimulated in this way go through three progressive stages—indifference, empathy, and integration.

The above survey is but a cross-section of the efforts that have been made to tap music's potentialities for the relief of psychological illness. It is not a dependable body of scientific knowledge, but is valuable in that it reveals the paramount needs which must be answered before the use of music as therapy can be developed. Many factors at present restrict and retard painstaking scientific research in this field.

IV. THE PROCEDURE EMPLOYED IN THIS STUDY

It should be evident from the foregoing historical data that the bulk of investigation of the application of music to the field of medicine has dealt with physiological phenomena. With the rapid growth of psychiatry as a dynamic medical specialty in recent years, the need became apparent to study the effects of applied music as an adjunct in the treatment of mental and emotional disorders. It was also recognized that a more definitive knowledge of the specific components of musical sounds should be presented systematically to the physician. The next step was the appraisal, on a controlled basis, of the specific effects or reactions evoked by known musical stimuli. An evaluation by qualified psychiatrists of the results obtained was considered essential. No previous studies were available in which an evaluation of both the music and its specific effects on the individual were measured and the results carefully recorded. The following practical plan was evolved.

A. *Preparation.* For the study, medical officers selected patients who, in their opinion, might respond favorably to music. Because of the lack of precedent, the problem of selection of cases was a highly individual matter with the physician. However, as the study progressed, the results obtained with certain types of disorders furnished a more uniform basis for selection. These patients were classified according to their predominant symptoms and their level of musical intelligence. The medical officers indicated on the prescription form not only the classification and diagnosis, but the mood and behavior changes that were desired.

Control cases were selected by the medical officers and were given identical treatment, except that they participated in no music sessions. The selection of controls was not an easy matter, since it is obvious that no two individuals suffer alike from the same mental illness. However, the criteria employed required that the subject for music sessions and the control case be of the same age group, race, color, sex, marital status, educational level, diagnosis, duration and severity of illness, and that both the subject and the control patient be under the observation of the same evaluating psychiatrist. Both patients received as nearly identical routine and special treatments as possible, with the additional factor of music sessions for the subject patient. In each instance of evaluation of the subjects and in all comments of the evaluating psychiatrist regarding the subject's response to music sessions, the control patient was used for comparison.

After selection and classification, each patient was then interviewed by the musical director to determine his "musical level," and was classified on the basis of this evaluation into one of four groups:

(1) Little or no familiarity with any music.
(2) Moderate familiarity with the simpler forms and expressions of musical composition.
(3) Educated musical taste or preference.
(4) Some degree of experience in participation.

This interview was conducted quite informally, but with definite direction and plan, so that when it was terminated, much pertinent data on the individual was available for use by the musician. This data included background information, the patient's home state and home town, his occupation, the nativity of his parents, the schools he had attended, the places he had been or wanted to visit, his branch of service, whether he had been overseas, and where.

Patients who fell into similar classifications, e.g. "Restlessness—educated musical taste—soothing therapy," or "Depression—moderate familiarity with music—stimulating therapy," were then assigned to small groups of from three to six members. A specific hour for the application of the music was arranged for each group and it met regularly at the same hour, five days a week. The duration of the treatment varied with the length of hospital residence, which averaged approximately three months but was frequently shorter.

The physical environment of the sessions was carefully ordered so as to predispose to a feeling of comfort, relaxation, and informality. The predominant color was subdued, but not somber. Chairs were upholstered, and there were facilities for reclining. The musical instruments were in the room and plainly visible, but were not given dominant position.

The piano was the instrument chiefly used, although violin, cello, harp, and solovox attachments were also employed. (An expanded study might well have included the comparative value of the various instruments.)

B. *Session Arrangement.* The musical-treatment sessions were divided into three parts:

1. The introductory or mood-determination and development period. Compositions selected with the aim of meeting the patients at the mood level that they brought to the sessions were played in order to establish a basic rapport between the patients and the music. Then, gradually and without any abrupt transition, music designed to develop the feeling tone prescribed by the medical officer was presented, simply and without obtrusion of the musician into the picture. The duration of this part of the session varied according to the patient's span of attention.

2. A brief interim period for establishing verbal

rapport between patients and musician, should the patients feel so inclined.

3. The period of patient participation. No direct invitation was issued, but the informality of the environment was conducive to this goal. Participation might be in the form of comments, queries, requests, or through humming, beating time, singing, whistling, or following the score. Whenever any patient showed especial desire or aptitude to express himself through music, he was encouraged to do so through arrangement for private instruction as well as through participation in group sessions. The results obtained in the limited number of cases that received such special instruction in music served to verify the expectation that this was an important aspect of therapy.

C. *Selection and Orientation of Personnel.* The plan called for continuity of musical personnel by musicians of merit and outstanding talent. Already qualified as professional musicians, and with a background of psychology or applied psychiatry, staff members were required, in addition, to conduct several sessions under observation and to attend weekly orientation periods supervised by a medical officer, where individual cases were discussed, procedures checked, and notes compared. These orientation periods were important for both musicians and medical officers, and much valuable information was exchanged.

The personality qualifications of the musicians

41

were of paramount importance. Especial difficulty was encountered in getting musicians to understand that the sessions were not intended to present opportunities for their own emotional expression or demonstration of the brilliance of their techniques, and that the spotlight was on the patients rather than on themselves. The extent and degree of their adaptability to the project was the most important consideration. Although they needed the skill of true artists, this alone was insufficient recommendation. The personality of the performer seemed to exert a definite influence upon the response of the patient. Furthermore, specific indoctrination in this work was a necessity.

D. *Classification of Music for the Purpose of the Study.* A committee of musicians and music educators prominent in national musical life devoted a great amount of time to placing all the musical selections used into a practical classification. The first was a technical classification based on the following criteria:

I. All music for use in hospitals should be first generally classified as follows:

    A. Music of solely rhythmic interest
    B. Music of solely harmonic interest
    C. Music of solely melodic interest

II. Of the first group (I), each subheading (A, B, C) should then be divided into two groups each (slow, fast) as follows:

    A. Music of modal nature—slow, fast
    B. Music of classic nature—slow, fast
    C. Music of romantic nature—slow, fast
    D. Music of impressionistic nature—slow, fast
    E. Music of modern modal nature—slow, fast

III. Of the second group (II), each subheading (A-E) should finally be subdivided as to key, length of piece, tempo, and character (program or absolute music).

In view of the fact that the doctor was limited in his prescription to requesting the emotional response, it became necessary for the musician to classify music further on the basis of the anticipated emotional effect. It is apparent that because of their extensive observation of audience reactions, professional musicians have a valuable background which equips them with a discriminating sense of selection. A survey of a cross-section of the vast quantity of available music resulted in the following simplified and practical classification:

I. Stimulating Music

    A. Mildly exciting—(Example—"The Bells of St. Mary's," by Adams)

    B. Joyous—(Example—"Spring Song," by Mendelssohn)

    C. Markedly exciting—(Example—"Bolero," by Ravel)

II. Sedative Music

A. Meditative—("Song to the Evening Star" from Tannhauser, by Wagner)

B. Soothing—("Lullaby," by Brahms)

C. Music with rhythmic flow—("Love's Old Sweet Song")

D. Music with poetic thought—"Clair de Lune," by Debussy)

E. Reveries—("The Rosary," by Nevin)

F. Depressing—("The Funeral March," by Chopin)

It is not to be presumed that selections calculated to be stimulating in a majority of instances might not have the reverse effect on occasion. This is probably best explained on the basis of associative memories on the part of the individual listener, and for this reason a basic acquaintance with the background and tendencies of the patients was necessary before individual selections could be intelligently chosen. With the most careful attention being given to individual difference, it was required that the repertoire of the musicians be submitted in advance. From these, selections were made that were applicable to the specific group to be treated.

Both medical and musical records were kept on all patients and on all sessions, using printed forms. The music records included the name of the composition, the key, the tempo, the instrument, and the comments of the musician on the patient's reactions. Between sessions the medical officers kept notes on patient's reactions, all data on each patient being correlated in frequent conferences between the staff musician and the physician.

V. CASE MATERIAL AND RESULTS

The following table summarizes the results obtained in fifty psychiatric cases.

TABLE I.

	Psychotic		Non-Psychotic		
	Stimulating Music	Sedative Music	Stimulating Music	Sedative Music	Totals (50 cases)
Favorable Response	4	8	9	16	37
No Response	—	1	2	3	6
Inconclusive	—	—	—	7	7
Average No. of Sessions	14.5	12.7	11.9	8	10.8

Unlike the known situation in psychotherapeutic efforts with psychotic patients, it is interesting to note that in the above tabulation the psychotic patient, once an initial rapport was established, attended a greater average number of sessions than the non-psychotic. The psychotic group attended an average of approximately 13 sessions, as compared to 10 for the non-psychotic group. Further comparison of the psychotic and non-psychotic group reveals that stimulating music had essentially the same effect in both groups, but sedative music had a decidedly better effect in the psychotic than in the non-psychotic

45

group. The nature of this experiment has been such that it would be highly speculative to attempt to account for these observed differences in the response of the psychotic and non-psychotic groups. Undoubtedly future investigation will reveal specific reasons for these observations. The encouraging fact is that, despite the numerous obstacles and handicaps which were encountered in this pioneer effort, in which 536 music sessions were conducted with 50 patients, a total of 37 patients, representing 74% of this series, showed sufficient improvement in their clinical condition for it to be noted and described by the evaluating psychiatrists.

A. *Illustrative Case Reports.*

Case One

A 22-year-old enlisted man was admitted to Walter Reed General Hospital in December, 1946, as a transfer from a nearby station hospital where he had been first hospitalized three weeks previously. He was eligible for a discharge on length of service at the time of his hospitalization but had become agitated, apprehensive, and suspicious, and hospitalization was recommended.

The family history revealed that the mother had suffered from cerebral thrombosis and that there were two siblings in good health. The family history was negative for neuropsychiatric determinants. The patient was born in New York City and was described as having been a rather moody child. He began school at the usual age and had completed high school and three years of university study at the time of his enlistment in the service early in 1944. The patient was married and there were no children.

On admission to Walter Reed he was agitated, depressed, and contemplating suicide. He appeared markedly hostile, clenching his fists, gritting his teeth, protruding his jaw, and grimacing. He was constantly mulling over numerous philosophical problems. He trusted no one. He believed that the lights in the ceiling were hypnotizing him, that there were microphones in the walls, and that everything that he did was being reported to the General Staff. He heard voices telling him to kill himself or else he would be killed.

Shortly after his admission, he was recommended for participation in the music program with specific prescription by the physician that he be given sedative and soothing music. He had had several years of instruction in the violin. He participated hesitantly at first and expressed considerable pessimism over the potentialities of music's being any help in his case. The staff musician recorded the following comments made by the patient during his first visit: "There is no use coming up here any more. If I could get out, it would be different." The following day he stated, "I used to have faith in human nature, but now I have faith in nothing and nobody except this," and he tapped his violin.

47

The patient continued to attend sessions regularly, and within a matter of days a noticeable change in his entire demeanor and outlook was observed. He appeared cheerful, conversed with others, listened attentively, and participated by playing on the violin. He preferred chamber music and ballads.

Following a particularly absorbing session, he commented, "I am wondering what I should do when I leave here. I feel so much better than when I first came." After a total of thirteen sessions, the patient was discharged from the hospital, and the following evaluation was recorded by his ward psychiatrist: "This patient was completely involved in personal and universal problems when he began taking part in the music program. During his period of participation, he gradually improved. Since he was always intensely interested in music, it is felt that participation in this program gave him an outlet that was lacking in other patient activities."

Case Two

A 30-year-old lieutenant, unmarried, was hospitalized at Walter Reed General Hospital in April, 1944, following two years of active combat duty in the Pacific theater and a long and depressing non-combat assignment at Teheran. He was born in New Orleans and was a graduate of Louisiana State University. At the time of his induction, he was holding a secretarial position.

On admission to the hospital, he was obsessed by

memories of his combat experiences. He was morbidly depressed and anxious. He was preoccupied
and emotionally dull, listless, and apathetic. On the
ward, he was seclusive and withdrawn, and one of his
most striking symptoms was a morbid fear that his
family or friends might see him in his present state.
The transfer diagnosis of schizophrenia, type un-
qualified, was made.

Although he had no musical background, music
therapy of a stimulating nature was prescribed in the
hope that music might furnish a means of establishing contact for psychotherapy. In interview, the
Music Director learned that Italian and French
operas were his favorite types of music and that he
was "allergic" to jazz.

During the first few music sessions, the patient
continued to brood in silence and showed no response
to either the music or musicians. During the fourth
or fifth session, the musician was successful in establishing verbal rapport with the patient. "I'm so mixed
up," he mumbled repeatedly, frowning and shaking
his head. From him the musician learned that he and
the girl to whom he had been engaged, but now re-
fused to see, had been very fond of dancing, and that
one of their favorite dance tunes was from "The
Student Prince." The patient could not remember
the specific selection. The musician sat down at the
piano and began the Romberg score. When she had
played a chord or two of "Deep in My Heart, Dear,"
the patient exclaimed, "That's it!" and began to re-

count pleasant experiences associated with the song and his pre-war years. This selection, and other similar ones, were played daily for the patient. He gradually gained confidence in the musician and in the situation, and talked with increasing ease about his home, his family, and his girl. This was carried back to his ward behavior, which showed progressive improvement. In a day or two he began to report to the music session neatly dressed in his uniform, in contrast to his previous slovenly, unkempt appearance. He consented to allow the musician to look up several of his friends and to notify them of his whereabouts. After a total of twelve sessions over a period of three weeks, he was discharged to his own care, and the psychiatrist's evaluation noted that his progress had been much more rapid than that of a control case who had had similar treatment with the exception of attending music sessions.

Case Three

A 43-year-old medical officer, married, was admitted to the hospital in a state of neurotic depression arising from a prolonged and disagreeable occupational assignment. His principal and striking symptoms on admission were a marked inability and uncertainty in expressing himself in conversation, and a morbid dread of leaving the building unattended (agoraphobia). Although he had an amateurish past interest in singing, his actual musical background had been largely appreciative rather than participatory. He attended twelve organized sessions of applied music and attended many hours in individual practice sessions over a period of three weeks. During this time he practised breath control and memorizing of music and lyrics. He soon showed willingness, and considerable ability, to entertain by singing before small groups. This in turn was followed by an increased sociability on the ward and attempts to take longer and longer walks by himself. He was discharged from the hospital as sufficiently recovered to justify return to his home, and before his departure he wrote the following letter to the Music Department, which in the opinion of the staff psychiatrist expressed an accurate picture of the role music had played in his recovery:

"This is to express my opinion as to the psychotherapeutic effects of music in a case of severe neurosis.

"Although I have always appreciated music and held the belief that music therapy would be beneficial to the neuro-psychiatric patient, especially in the stage of convalescence, I now am convinced from actual experience that these psycho-therapeutic effects are more real than theoretical.

"This form of therapy gives the patient an opportunity of 'acting out' his emotions instead of submerging them deeply into the sub-conscious strata. Because of the numerous items that one must think of while studying music, i.e. breath control, memorizing words, music, etc., the patient has very little time to dwell on his deleterious complexes and thus gradually establishes a tendency to become more extrovert.

"Another important factor is the relaxation that the patient experiences, thereby increasing his self-confidence. Music as it is applied and taught daily to patients at the Walter Reed Hospital also affords an opportunity to socialize with others in an area of mutual interest.

"All the above mentioned factors tend to eventually (and I believe successfully) remove the feeling of inadequacy which is constantly grasping the neurotic patient."

VI.   SUMMARY

With the recognition that it was possible to design an experiment which would evaluate the effects of music in the treatment of neuropsychiatric disorders, this study was begun at Walter Reed General Hospital in March, 1944, by the Department of Applied Music. The ever-increasing incidence of neuro-psychiatric disorders requiring hospital care indicated the need for adjuncts to psychiatric treatment. Historically, it is noted that music as applied to the mentally ill possesses certain "healing constituents" which have never been satisfactorily evaluated. A plan and method of study utilizing the close co-operation of qualified musicians and psychiatrists was devised and rigidly adhered to in so far as practical circumstances would permit. This plan provided three major innovations: (a) music was never administered to a patient without a physician's prescription specifying the predominant symptoms and the desired behavior change; (b) music was specially selected to produce the predetermined change, utilizing wherever possible pertinent background facts from the patient's associative experiences; and (c) patients were grouped homogeneously according to both medical and musical needs. In its early phases, this study faced many preconceived attitudes. Naturally, each individual concerned with the study believed that he was entitled to an opinion regarding its validity.

As the plan was developed over a period of years, it became apparent to the majority of those concerned that the "x-factor" in music as applied to mental patients, despite the limited understanding of it, manifested itself in recurring patterns of success that could be systematically arranged and recorded as clinical observations by psychiatrists. Application of the procedure and forms devised for this study yielded certain favorable clinical results, in our opinion, which are presented in tabular form and illustrated by three case abstracts.

VII. CONCLUSIONS

1. A systematic and skillful application of music in neuro-psychiatric hospitals has proved to be a definite adjunct to the psychiatric regime.

2. The therapeutic value of music has not yet been exploited to the fullest advantage.

3. There is an urgent need for future controlled studies and further development of the procedures and methods devised in this study over a period of three and one-half years.

REFERENCES

1. Analects of Confucius, Arthur Waley.
2. Bryan, W. L., and Bryan, C. L.: The Republic of Plato. New York, 1898, p. 316.
3. Le Massena, C. E.: Latent Therapeutic Values in Music. Musician, Oct. 1940, p. 142.
4. Burton, Robert: The Anatomy of Melancholy. London, 1621.
5. Iamblickus: Life of Pythagoras.
6. Dogiel, J.: Über den Einfluss der Musik auf den Blutkreislauf. Arch. F. Physiol., 1880, 416-428.
7. Schoen, Max: The Psychology of Music. Ronald Press, N. Y., 1940, p. 106.
8. Hyde, Ida H.: Effects of Music upon Electrocardiograms and Blood Pressure. Journal of Experimental Psychology, 1924, 7: 213-224.
9. Schoen, Max: The Psychology of Music. Ronald Press, N. Y., 1940, p. 106-107.
10. Diserens, Charles M., and Fine, H.: A Psychology of Music: The Influence of Music on Behavior. College of Music, Cincinnati, Ohio, 1939, p. 239.
11. Foster, J. C. and Gamble, E. A.: The Effect of Music on Thoracic Breathing. American J. Psychology, 1906, 17: 406-414.
12. Murtfeldt, E. W.: How Music Heals the Sick. Popular Science Monthly, Oct. 1937.
13. Diserens, Charles M., and Fine, H.: A Psychology of Music: The Influence of Music on Behavior. College of Music, Cincinnati, Ohio, 1939, p. 214.

14. Schoen Max, Ph.D.: Doctoring with Music. Etude, March 1942, p. 166.
15. "Music May be Valuable in Mental Treatment," Hygeia, July 1935, p. 664.
16. Diserens, Charles M., and Fine, H.: A Psychology of Music: The Influence of Music on Behavior. College of Music, Cincinnati, Ohio, 1930, p. 166-168.
17. Ibid.
18. Ibid.
19. Megliorino, G.: Emotions Arising from Music. 1938, pp. 58, 485-501.
20. Diserens, Charles M., and Fine, H.: A Psychology of Music: The Influence of Music on Behavior. College of Music, Cincinnati, Ohio, 1939, p. 156.
21. Schoen, Max, Ph.D.: The Psychology of Music. Ronald Press, 1940, p. 89.
22. Ibid., pp. 97-98.
23. Ibid., pp. 102.
24. Ibid., pp. 102-103.
25. Montani, Angelo: Psychoanalysis of Music. Psychoanalytic Review, April 1945, Vol. 32, pp. 225-227.
26. Tilly, Margaret: The Psychoanalytical Approach to the Masculine and Feminine Principles in Music. Am. J. Psycho., 103: 477-83, Jan. 1947.
27. Pfeifer, Sigmund: Problems of the Psychology of Music in the Light of Psychoanalysis. Abstracted in the Inter. Journal of Psycho-analysis, 1922.
28. Van Der Chijs, A.: An Attempt to Apply Objective Psychoanalysis to Musical Composition. Abstracted in Inter. J. Psycho-analysis. IV: 379-80, 1923.
29. Coriat, Isadore H.: Some Aspects of a Psychoanalytic Interpretation of Music. Psychoanalytic Review, 32: 408-18, Oct. 1945.
30. Antrim, Doron K.: Music Therapy. The Musical Quarterly, 30: 416, Oct. 1944.
31. Ibid., p. 417.
32. Altschuler, Ira., One Year's Experience with Group Psychotherapy. Mental Hygiene, 24: 190-96, April 1940.

THE CAPURSO STUDY

by Alexander Capurso, Ph.D.*

> Music produces like effects on the mind
> as good medicine on the body.—*Mirandola.*

HOW THESE TESTS WERE MADE

The experiments, the results of which are presented here, were undertaken in March, 1948, by the author, who was then head of the Music Department at the University of Kentucky, with the assistance of a grant by the Music Research Foundation. The study was put in progress immediately at the University of Kentucky, with the assistance of faculty members and students at Transylvania College, Lexington, Kentucky. Following this, the author assumed the directorship of the School of Music of Syracuse University, and the experiments were continued there.

The first step was to secure from one hundred and thirty-four college and high school music instructors the names of certain compositions which in their opinion were associated with the mood categories specified below. One hundred and five of the selections mentioned most frequently by the experts were

* Director of the School of Music, Syracuse University.

then chosen to be used in listening tests with 1075 non-musical students.

The experts had certain restrictions on their choice of selections in that they were limited (1) to instrumental music, solo or ensemble, and (2) to vocal music presented in languages other than English. A further limit was imposed regarding (3) the length of the selections, making it necessary to choose short works or designated sections from the larger forms, with most selections lasting between three and four minutes.

The listening tests were conducted under uniform conditions, utilizing distraction-proof listening rooms, high fidelity phonographs, unworn and undamaged records, and reasonable control of sound volume.

Some selections were tested by as many as 700 students, and none was tested by less than 100 students.

At the completion of the playing of each selection the students were asked to indicate which of the following six categories they would select to describe the effects of the composition:

A: Happy, Gay, Joyous, Stimulating, Triumphant
B: Agitated, Restless, Irritating
C: Nostalgic, Sentimental, Soothing, Meditative, Relaxing
D: Prayerful, Reverent

E: Sad, Melancholy, Grieving, Depressing, Lonely
F: Eerie, Weird, Grotesque

In order to avoid as far as possible "serial effects" of the experiment, that is, any influence on the subjects' responses that might be effected by the mood caused by the preceding selection, at least two minutes were permitted to elapse between the playing of each selection. During this time, the subjects were encouraged to engage in conversation on any topic not related to music listening. Furthermore, before the playing of subsequent selections, the subjects were asked if they were ready to listen to the next work, and whether they could do so without being influenced by the prior work.

Although five of the six categories of mood qualities include more than two suggestive mood adjectives, it is conceded that the grouping of the moods might have had some slight suggestive effect upon the listeners. In the study described here, the author has as yet been unable to control this factor. Further studies, however, are in the process of being structured to increase the value of the individual subjective responses.

It is to be hoped that the present study, in spite of its limitations, will serve as a useful paradigm for later investigations and that it will encourage workers in the field of music research to engage in further exploration of this important issue.

THE CAPURSO STUDY

MUSICAL SELECTIONS LISTED ACCORDING TO
THEIR MOOD CATEGORY

The musical selections of the following list are suitable not only for creating a desired emotional effect on the listener, but also for selecting suitable *background music for radio and television programs.*

It should be kept in mind that some of the selections listed were placed in different categories by different people. The actual percentage of the people that agreed on any one category is given in the column headed "Listener Agreement." For instance, *The Stars and Stripes Forever* was placed in category A (Stimulating) by 93% of the people listening to it, while the other 7% of the listeners put it in category B (Agitating). Thus, when a selection is placed in a certain category, it means that a greater percentage of persons associate it with that mood rather than with the other moods.

The number given in the column headed "Emotional Strength" is the average intensity of the reaction of the listeners to the selection. Such stirring pieces as *The Stars and Stripes Forever* or the *Prelude to Act III of Lohengrin* rated 90% emotional strength, while selections such as *Humoresque* and *Moto Perpetuo,* though in the same mood category, were able to raise much less emotion. Emotional strength was measured by asking the subject to express verbally the intensity of his own response in accordance with a point scale, which was then transformed into a percentage figure.

It will be noticed that of the 105 selections only those 61 pieces which provide listener agreement of 50% or more are presented in the following table. Appendix B contains the complete list of 105 selections, together with the percentage of agreement in each mood category. Appendix B also lists the original records used in this investigation. Since some of the records originally used in the study may not be readily available to the reader, the following table presents a number of other performances of the same musical selection.* It should be understood, however, that the results of the study are, strictly speaking, directly relevant only to those recordings actually used in it, and these are given in Appendix B. Anyone interested in further details of this study is invited to write to Music Research Foundation for them.

Finally, it should be mentioned that the comments listed under each selection in the following table were provided by Miss Selma Kurz, and are not to be interpreted as findings of the study itself.

* The listings were taken from the current catalogs of the available 33⅓ and 45 RPM recordings of the pieces of music.
The capital letters in parentheses, following the names of the performers, indicate the maker of the record: (V) for Victor, (C) for Columbia, and (D) for Decca.

MOOD CATEGORY A (HAPPY, GAY, JOYOUS, STIMULATING, TRIUMPHANT)

TITLE AND COMPOSER	EMOTIONAL STRENGTH	LISTENER AGREEMENT	PERFORMER
1. Stars and Stripes Forever March—Sousa	94%	93%	Ralph Flanagan (V) Toscanini, N.B.C. Symphony Orchestra (V) Horowitz (piano) (V) J. Colling, Hollywood American Legion Band (D)

Comment: A proud, exciting march. Its angular melody, moving briskly in lively, emphatic rhythms, is articulated within a flashy orchestration.

TITLE AND COMPOSER	EMOTIONAL STRENGTH	LISTENER AGREEMENT	PERFORMER
2. William Tell: Overture (Finale)—Rossini	84%	91.3%	Toscanini, N.B.C. Symphony Orchestra (V) Kostelanetz and Orchestra (C) Van Kempen, Berlin Philharmonic Orch. (D)

Comment: Zestful melodies, interspersed with more pensive moments.

TITLE AND COMPOSER	EMOTIONAL STRENGTH	LISTENER AGREEMENT	PERFORMER
3. Symphony No. 5 (Second movement)— Shostakovitch	70%	88.7%	Rodzinski, Cleveland Orchestra (C)

Comment: Energetic, of great rhythmic vitality, replete with pungent, often dissonant harmony.

MOOD CATEGORY A (HAPPY, GAY, JOYOUS, STIMULATING, TRIUMPHANT)–(*Continued*)

TITLE AND COMPOSER	EMOTIONAL STRENGTH	LISTENER AGREEMENT	PERFORMER
4. Symphony No. 35 in D major ("Haffner") (Fourth movement)—Mozart	84%	88.6%	Reiner, Pittsburgh Symphony Orchestra (C)
			Karajan, Italian Radio Symphony Orchestra of Turin (D)
			Toscanini, N.B.C. Symphony Orchestra (V)

Comment: A glittering, exquisitely designed Presto movement, delicate and merry. It begins softly in the strings, grows in vigor as it moves, without losing clarity, to the end.

TITLE AND COMPOSER	EMOTIONAL STRENGTH	LISTENER AGREEMENT	PERFORMER
5. Washington Post March—Sousa	90%	88.2%	Cities Service Band (V)
			J. Colling, Hollywood American Legion Band (D)

TITLE AND COMPOSER	EMOTIONAL STRENGTH	LISTENER AGREEMENT	PERFORMER
6. Aida: Grand March—Verdi	84%	87.3%	Boston Pops Orchestra (V)

Comment: A victorious processional march. The broad, stately melody, with its brilliant, trumpet-like flourishes, is supported by a martial rhythmic pattern retained throughout.

TITLE AND COMPOSER	EMOTIONAL STRENGTH	LISTENER AGREEMENT	PERFORMER
7. Lohengrin: Prelude to Act III—Wagner	96%	87.2%	Reiner, with the Pittsburgh Symphony Orchestra (C)

Comment: Exultant music. A brilliant fanfare, rising melodically through the sturdy G major chord, heralds the jubilant theme. Later, the fanfare motif, to which the unstable interval of the Seventh has been added, appears alone in the bassoons and horns; the strings maintain a dynamic rhythmic pattern beneath it. Together, they create an effect of gradually mounting excitement, climaxed by the re-entrance of the jubilant theme. A quiet interlude follows, in which gentle winds and lyric strings treat the fanfare motif in a stately, dance-like rhythm. Ultimately, the jubilant theme returns in its full sweep, driving on to a brilliantly sonorous conclusion.

TITLE AND COMPOSER	EMOTIONAL STRENGTH	LISTENER AGREEMENT	PERFORMER
8. Symphony No. 7 in A major (Third movement)—Beethoven	84%	86%	Ormandy with the Philadelphia Orchestra (C)
			Walter, N.Y. Phil.-Symphony Orchestra (C)
			Munch, Boston Symphony Orchestra (V)
			Toscanini, N.Y. Phil.-Symphony Orchestra (V)

Comment: Rapid and irresistibly vigorous. An exuberant display of brilliance, although more solemn moods intrude occasionally. Toward the end, the hymn-like melodic pattern prevails, only to be drowned out by five powerful chords.

MOOD CATEGORY A (HAPPY, GAY, JOYOUS, STIMULATING, TRIUMPHANT)—(Continued)

TITLE AND COMPOSER	EMOTIONAL STRENGTH	LISTENER AGREEMENT	PERFORMER
9. Moto Perpetuo—Paganini	68%	81.9%	Boston Pops Orchestra (V)

Comment: Breathless and breathtaking in speed and brilliance, yet a tuneful and well-balanced array of passages, famous as a bravura piece.

TITLE AND COMPOSER	EMOTIONAL STRENGTH	LISTENER AGREEMENT	PERFORMER
10. Piano Concerto No. 5 (Third movement) —Beethoven	76%	79.1%	Serkin, piano, with The Philadelphia Orchestra (C)
			Schnabel, piano, with Chicago Symphony Orch. (V)

Comment: Lively, in parts gruffly good natured. A catchy main tune imbedded in a full-blooded piano-orchestra combination.

TITLE AND COMPOSER	EMOTIONAL STRENGTH	LISTENER AGREEMENT	PERFORMER
11. Incidental Music to A Midsummer Night's Dream: Scherzo—Mendelssohn	82%	77.8%	Fricsay, Berlin Philharmonic Orchestra (D)
			Toscanini, N.B.C. Symphony Orchestra (V)

Comment: Elfin humor, bright as dancing fireflies, sparkles through a delicate orchestration. Dance-like motifs in rapid, vivacious rhythm set up the ceaseless momentum maintained throughout the piece. With exquisite spacing, so that the individual characteristics of the instruments are retained although sonority is increased, strings present their own theme, a driving motif in perpetual motion. The sprightly interplay of themes and instruments ultimately carries the music into the highest register. There is a flicker of the dance motif in the winds, then all vanish.

TITLE AND COMPOSER	EMOTIONAL STRENGTH	LISTENER AGREEMENT	PERFORMER
12. Symphony No. 5 (Fourth movement)—Beethoven	76%	76.7%	Walter, N.Y. Philharmonic Symphony Orchestra (C) Koussevitzky, Boston Symphony Orchestra (V)

Comment: A broad, resplendent melody, powerful and sonorous, full of lively contrapuntal accents, and yet in places also indicating the echoes of those mysterious premonitions which characterize the first movement. It ends in a spirit of reconciliation and that form of calm control which covers a smoldering fire always ready to rise into flames again.

TITLE AND COMPOSER	EMOTIONAL STRENGTH	LISTENER AGREEMENT	PERFORMER
13. Polonaise in A flat major, Opus 53, No. 6—Chopin	84%	73.5%	Iturbi (piano) (V)

Comment: Vigorous dance rhythms proclaim in terms of fierce masculinity the deep-rooted pride of country and the profound devotion felt by a true patriot.

TITLE AND COMPOSER	EMOTIONAL STRENGTH	LISTENER AGREEMENT	PERFORMER
14. Symphony No. 41 in C ("Jupiter") (Fourth movement)—Mozart	74%	73.1%	Walter, N.Y. Phil.-Symphony Orchestra (C) Beecham, Royal Philharmonic Orchestra (C) Toscanini, N.B.C. Symphony Orchestra (V)

Comment: One melody goes through the entire movement. Its theme is broad and majestic, and evolves into a fugue of unexcelled inner strength.

MOOD CATEGORY A (HAPPY, GAY, JOYOUS, STIMULATING, TRIUMPHANT)—*(Continued)*

TITLE AND COMPOSER	EMOTIONAL STRENGTH	LISTENER AGREEMENT	PERFORMER
15. Humoresque—Dvorak	62%	73%	Kreisler (Violin) (V)
			Elman and Mittman (Violin and Piano) (V)

Comment: Blithe, enticing music of "velvet" texture, based on Czech folk-song motives; it breathes homey warmth and quiet, serene enjoyment.

TITLE AND COMPOSER	EMOTIONAL STRENGTH	LISTENER AGREEMENT	PERFORMER
16. Symphony No. 5 in C minor (First movement)—Beethoven	64%	70.4%	Walter, N.Y. Phil.-Symphony Orchestra (C)
			Koussevitzky, Boston Symphony Orchestra (V)

Comment: The movement, which begins with its characteristic ominous rhythm, reflects a mood of anxious expectation. The mind is filled with dark presentiments that constrict the chest and take the breath away. Until, after a motif elaboration of unsurpassed strength and beauty, a glorious resolution like a fiery torch brightens the darkness of the night.

TITLE AND COMPOSER	EMOTIONAL STRENGTH	LISTENER AGREEMENT	PERFORMER
17. Symphony No. 4 in E minor (First movement)—Brahms	60%	69.6%	de Sabata, Berlin Philharmonic Orchestra (D)
			Walter, N.Y. Phil.-Symphony Orchestra (C)

Comment: Crisp, fluid, in places deeply lyrical. The movement gains in force and persuasiveness, and ends on a note of quiet optimism.

TITLE AND COMPOSER	EMOTIONAL STRENGTH	LISTENER AGREEMENT	PERFORMER
			Ormandy, Philadelphia Orchestra (C)
			Munch, Boston Symphony Orchestra (V)
18. Turkey in the Straw—Folk Tune	84%	69.1%	Guy Lombardo (D)
			Boston Pops Orchestra (V)

Comment: Hardy music, delightful and contagious. Its keen and jaunty rhythms, its frolicsome, happy melody makes it a medium of fun and joy.

TITLE AND COMPOSER	EMOTIONAL STRENGTH	LISTENER AGREEMENT	PERFORMER
19. The Messiah: Hallelujah Chorus—Handel	82%	68.7%	Luton Choral Society (V)
			Robert Shaw Chorale (V)
			Huddersfield Choral Society, Liverpool Orchestra, Sir Malcolm Sargent (C)
			Chorus, Royal Philharmonic Orchestra, Sir Thomas Beecham (V)

Comment: Regally triumphant music. Above a vigorous rhythmic pattern in the bass strings, the incisive Hallelujah theme is articulated with massive strength. It is an angular melody in an energetic rhythm, based upon the intervals of a simple major chord. Later it appears as a sharp accent to an arched melodic line. The music concludes as all of the melodic lines blend into a paean of triumph.

MOOD CATEGORY A (HAPPY, GAY, JOYOUS, STIMULATING, TRIUMPHANT) —*(Continued)*

TITLE AND COMPOSER	EMOTIONAL STRENGTH	LISTENER AGREEMENT	PERFORMER
20. Etude in G flat major, Opus 10, No. 5— Chopin	66%	65.5%	Brailowsky (piano) (V)
			Byron Janis (piano) (V)
			First Piano Quartet (V)

Comment: A masterful "black-key" etude. While the right hand remains on the five black keys throughout the composition, the left hand displays much more freedom of movement in the pearly tonal mosaic as well as in the distribution of dynamic accents. A lively piece; almost "salon music."

TITLE AND COMPOSER	EMOTIONAL STRENGTH	LISTENER AGREEMENT	PERFORMER
21. Ritual Fire Dance (From "El Amor Brujo")—DeFalla	74%	63.9%	Levant (piano) (C)
			Kostelanetz and Orchestra (C)
			Iturbi (piano) (V)
			Rubinstein (piano) (V)
			Whittemore and Lowe (Duo-pianists) (V)
			Boston Pops Orchestra (V)

Comment: Diabolic trills introduce a primitive, darkly exotic driving rhythmic pattern persisting throughout. Above it, the melody moves in grotesque patterns of minor second intervals, in a fiercely surging, barbaric tom-tom rhythm. As the drive, the tempo and the sonorities are intensified, the dance reaches a shattering climax.

TITLE AND COMPOSER	EMOTIONAL STRENGTH	LISTENER AGREEMENT	PERFORMER
22. An American in Paris—Gershwin	68%	61.4%	Bernstein, R.C.A. Victor Orchestra (V)
			Paul Whiteman's Orchestra (D)
			Sandford, Kingsway Symphony Orchestra (D)
			Templeton (piano), with Kostelanetz and Orch. (C)
			Rodzinski, N.Y. Phil.-Symphony Orchestra (C)
			Gershwin (piano) with Paul Whiteman's Orch. (V)

Comment: A rhapsodic ballet in French style, portraying the impression of an American visitor absorbing the French atmosphere. The opening section is followed by a nostalgic blues with a strong rhythmic undercurrent. The blues character changes toward the end when the spirit of the music returns to the bubbling exuberance of the opening.

TITLE AND COMPOSER	EMOTIONAL STRENGTH	LISTENER AGREEMENT	PERFORMER
23. Mefisto Waltz—Liszt	70%	60%	Rodzinski, N.Y. Phil.-Symphony Orchestra (C)
			Kapell (piano) (V)

Comment: Brilliant and garish; the music makes no great demands on concentration, the listener is impressed by its superlative orchestration of a wild, sensual mood. Toward the end the music becomes subdued and fades out.

MOOD CATEGORY A (HAPPY, GAY, JOYOUS, STIMULATING, TRIUMPHANT)—*(Continued)*

TITLE AND COMPOSER	EMOTIONAL STRENGTH	LISTENER AGREEMENT	PERFORMER
24. Polka from "The Age of Gold" Ballet—Shostakovitch	78%	60%	Whittemore and Lowe (Duo Pianists) (V) Kurtz, N.Y. Phil.-Symphony Orchestra (C)

Comment: In the orchestral version, the xylophone is prominent. It carries the main theme. The second theme is given to the saxophone. Other solos: violin, flute, and horn. The clarinet executes a cadenza.

TITLE AND COMPOSER	EMOTIONAL STRENGTH	LISTENER AGREEMENT	PERFORMER
25. Blue Danube Waltz—Strauss	86%	56.9%	Kostelanetz and Orchestra (C) Stokowski and Orchestra (V) Ferenc Fricsay and Berlin Philharmonic (D)

Comment: Vigorous triple meter, gay and buoyant melodies based upon simple chords, with clear, richly orchestrated harmonic accompaniment. Lilting rhythmic pattern enhances the danceable quality of the music.

TITLE AND COMPOSER	EMOTIONAL STRENGTH	LISTENER AGREEMENT	PERFORMER
26. Golliwog's Cake Walk (from the "Children's Corner Suite")—Debussy	58%	54.5%	Kapell (piano) (V)

Comment: Humorous, slightly grotesque, jolly music.

TITLE AND COMPOSER	EMOTIONAL STRENGTH	LISTENER AGREEMENT	PERFORMER
27. Fantaisie-Impromptu, Opus 66—Chopin	76%	53.1%	Iturbi (piano) (V) First Piano Quartet (V) Eileen Joyce (piano) (D)

Comment: The C-sharp minor work begins with a lively, light weight Allegro agitato which is followed by a moderato cantabile. The latter is known to many Americans as the background of the popular song, "I'm Always Chasing Rainbows." The agitated part then reappears, to be followed by a Coda in which motives of the cantabile are repeated.

TITLE AND COMPOSER	EMOTIONAL STRENGTH	LISTENER AGREEMENT	PERFORMER
28. Die Walküre: Ride of the Valkyries—Wagner	82%	50%	Toscanini, N.B.C. Symphony Orchestra (V) Reiner, Pittsburgh Symphony Orchestra (C)

Comment: Hoof-beats ring and comets crash as winged steeds career Wotan's daughters across the universe. The music depicts the exaltation and daring of the Teutonic amazons in a spectacular evocation of pagan imagery.

MOOD CATEGORY B (AGITATED, RESTLESS, IRRITATING)

TITLE AND COMPOSER	EMOTIONAL STRENGTH	LISTENER AGREEMENT	PERFORMER
1. Symphonie Fantastique (Finale)—Berlioz	66%	58.1%	Ormandy, Philadelphia Orchestra (C) Monteux, San Francisco Symphony Orchestra (V)

Comment: Witches' Sabbath. The composer says about this movement: "Now he (the hero, a 'morbidly inclined young musician') sees himself in a frightful company—ghosts, magicians, monsters—who have come to mourn over him. Briefly he hears the melody (representing his beloved), but it is transformed, vulgar, grotesque... She joins him in the infernal orgy."

TITLE AND COMPOSER	EMOTIONAL STRENGTH	LISTENER AGREEMENT	PERFORMER
2. Baba-Yaga—Liadov	54%	57.4%	

TITLE AND COMPOSER	EMOTIONAL STRENGTH	LISTENER AGREEMENT	PERFORMER
3. Flight of the Bumble-Bee—Rimsky-Korsakoff	64%	55.3%	Heifetz (violin) (V)

Comment: Above the energetic rhythmic accompaniment, the whirring melody of the virtuoso violin speeds brilliantly in consecutive minor second intervals, soaring and swooping with carefree abandon.

TITLE AND COMPOSER	EMOTIONAL STRENGTH	LISTENER AGREEMENT	PERFORMER
4. Sonata, Opus 35, for Piano (First movement)—Chopin	62%	53.7%	Casadesus (C) Horowitz (V) Rubinstein (V)

Comment: Turbulent passions and fantasies struggling to possess the soul of the hypersensitive and imaginative composer. Slow, pleading octaves precede a trembling dissonant chord, quickly resolved, but still apprehensive. Suddenly, a nervous, driving rhythmic pattern is introduced, accentuated by a tormented melodic motif based on the interval of the minor third. A tender theme in broad harmonies later appears, followed by a series of ringing chords. Dramatically, the tormented theme interrupts; throughout the composition it conflicts with the confident elements of the music. In a final, defiant gesture, it returns, then evolves into brilliant chords ending the movement.

73

MOOD CATEGORY C (NOSTALGIC, SENTIMENTAL, SOOTHING, MEDITATIVE, RELAXING)

TITLE AND COMPOSER	EMOTIONAL STRENGTH	LISTENER AGREEMENT	PERFORMER
1. Rhapsody in Blue (Ballad Theme)—Gershwin	92%	78.7%	Bargy, piano, with Whiteman Orchestra (D)
			Sandford, piano, with Camarata and the Kingsway Symphony Orchestra (D)
			Levant, piano, with Ormandy and the Philadelphia Orchestra (C)
			Templeton, piano, with Kostelanetz and Orchestra (C)
			Gershwin, piano, with Whiteman Orchestra (V)
			Jose and Amparo Iturbi, two pianos (V)
			Paul Whiteman's Orchestra (D)

Comment: The middle section of the work, based primarily on Negro spirituals; a broad, sentimental melody, like a blues tune.

TITLE AND COMPOSER	EMOTIONAL STRENGTH	LISTENER AGREEMENT	PERFORMER
2. Liebestraum No. 3—Liszt	92%	76.6%	First Piano Quartet (V)
			Iturbi, piano (V)
			Boston Pops Orchestra (V)

Comment: Intense, yet serene in expression; a lingering, romantic melody.

TITLE AND COMPOSER	EMOTIONAL STRENGTH	LISTENER AGREEMENT	PERFORMER
3. Serenade—Schubert	72%	70.6%	Nelson Eddy (C)
			Lotte Lehmann (V)
			Melton (V)
			Grace Moore (D)

Comment: A love song, pleading, eloquent, tender.

TITLE AND COMPOSER	EMOTIONAL STRENGTH	LISTENER AGREEMENT	PERFORMER
4. Clair de Lune—Debussy	78%	69.4%	Stokowski and Orchestra (V)
			Iturbi (piano) (V)

Comment: A tender orchestral work "spun of moonlight and stardust" into a rich tapestry of glittering harmonic designs.

TITLE AND COMPOSER	EMOTIONAL STRENGTH	LISTENER AGREEMENT	PERFORMER
5. Lullaby—Brahms	88%	69.2%	Blanche Thebom (V)
			Robert Shaw Chorale (V)

Comment: Blissful serenity. The melody moves in a triple meter with a gentle, rocking rhythm. Simple chords with stable intervals impart a sense of peace and security.

MOOD CATEGORY C (NOSTALGIC, SENTIMENTAL, SOOTHING, MEDITATIVE, RELAXING) —(*Continued*)

TITLE AND COMPOSER	EMOTIONAL STRENGTH	LISTENER AGREEMENT	PERFORMER
6. Pavane pour une enfante défunte—Ravel	72%	68%	Kostelanetz and Orchestra (C)

Comment: High and tenuous melody, moving with grace and dignity. Only when it suddenly sags do we notice the tension that was immanent in the composition from the onset.

TITLE AND COMPOSER	EMOTIONAL STRENGTH	LISTENER AGREEMENT	PERFORMER
7. The Swan, from Carnival of the Animals —Saint-Saëns	74%	66.7%	Primrose (viola) (V) Heifetz and Bay (violin and piano) (D)

Comment: Broad, lush music, stimulating, yet also meditative. It conveys the image of a swan sailing swiftly, majestically across a shadowed pool and vanishing in the darkness.

TITLE AND COMPOSER	EMOTIONAL STRENGTH	LISTENER AGREEMENT	PERFORMER
8. Waltz of the Flowers from the Nut-cracker Suite—Tschaikowsky	72%	58.9%	Irving, Royal Opera Orchestra (D) Kostelanetz and Orchestra (C) Rodzinski and N.Y. Philharmonic-Symphony Orchestra (C)

Comment: A graceful and subtle dance from a famous orchestral work. It captures the gossamer texture and delicate shading of frail springtime blossoms.

TITLE AND COMPOSER	EMOTIONAL STRENGTH	LISTENER AGREEMENT	PERFORMER
9. Air on the G String—Bach	76%	56.9%	Ormandy and the Philadelphia Orchestra (V) Stokowski and Orchestra (V)

Comment: A broad stream of melody maintaining throughout the mood of quiet bliss, rising to supreme heights of expression.

TITLE AND COMPOSER	EMOTIONAL STRENGTH	LISTENER AGREEMENT	PERFORMER
10. Score from the motion picture "The Lost Weekend"—Rozsa	72%	54.2%	Toscanini, N.B.C. Symphony Orchestra (V)

Comment: This is slow-moving music in a sentimental vein, dramatically intensified by an undercurrent of melancholy longing.

TITLE AND COMPOSER	EMOTIONAL STRENGTH	LISTENER AGREEMENT	PERFORMER
11. "Moonlight" Sonata (First movement)—Beethoven	74%	54.1%	Serkin (piano) (C) Horowitz (piano) (V) First Piano Quartet (V) Kempff (piano) (D)

Comment: Transcendental music, radiant in its clarity and daring in its dramatic power, set against a background of darkness and mystery.

MOOD CATEGORY C (NOSTALGIC, SENTIMENTAL, SOOTHING, MEDITATIVE, RELAXING) —(Continued)

TITLE AND COMPOSER	EMOTIONAL STRENGTH	LISTENER AGREEMENT	PERFORMER
12. Symphony No. 8 ("Unfinished") (First movement)—Schubert	70%	52.4%	Beecham, Royal Philharmonic Orchestra (C)
			Walter, Philadelphia Orchestra (C)
			Koussevitzky, Boston Symphony Orchestra (V)
			Toscanini, N.B.C. Symphony Orchestra (V)

Comment: Somber in mood, the movement soon unfolds to an undulating passionate melody with many climaxes and varying intensity, until it ends with a group of four powerful chords reflecting the deep, restrained emotions.

TITLE AND COMPOSER	EMOTIONAL STRENGTH	LISTENER AGREEMENT	PERFORMER
13. Scheherazade Suite (First movement)— Rimsky-Korsakoff	52%	52%	Ormandy, Philadelphia Orchestra (C)
			Monteux, San Francisco Symphony Orchestra (V)

Comment: The work opens with a stern and menacing theme (Sultan motive) which is followed by the lilting melody of Scheherazade (violin solo). Several colorful episodes unfold before our ears and eyes, and the work ends with an apotheosis of the Bronze Warrior.

TITLE AND COMPOSER	EMOTIONAL STRENGTH	LISTENER AGREEMENT	PERFORMER
14. Symphony No. 6, Opus 68 (Second movement: "By the Brook")—Beethoven	54%	50%	Walter, Philadelphia Orchestra (C)

Comment: Above the steady, gentle murmuring of limpid strings in a dance-like triple meter, there flows an abundantly warm melody. As it expands in sonority and breadth, strings and placid winds merge in a wholehearted abandonment to tenderness. A touch of whimsy at the end of the movement inspires the brief calls of nightingale, quail, and cuckoo.

79

MOOD CATEGORY D (PRAYERFUL, REVERENT)

TITLE AND COMPOSER	EMOTIONAL STRENGTH	LISTENER AGREEMENT	PERFORMER
1. Organ Choral No. 1—Franck	72%	78.7%	

Comment: A mood of serenity is created by the even-flowing blend of rich harmonies.

TITLE AND COMPOSER	EMOTIONAL STRENGTH	LISTENER AGREEMENT	PERFORMER
2. Mass in B minor (Crucifixus)—Bach	66%	71.1%	Robert Shaw with RCA Victor Chorale (V)

Comment: In this movement the music uses the rhythms of an ancient Spanish dance, the passacaglia, to express the spirit of extreme grief. The ground bass descends chromatically for the duration of four bars, a motion repeated not less than thirteen times. The voices enter one by one, uttering with amazed horror the word "crucifixus"—He was crucified. The last bars are sung by the chorus alone, while the basses continue with a figure resembling sobbing. The descending line of "sepultus est"–He was buried– reflects the mood of the burial. The movement ends with a sudden switch from E minor to G major in the final chord.

TITLE AND COMPOSER	EMOTIONAL STRENGTH	LISTENER AGREEMENT	PERFORMER
3. Xerxes: Largo—Handel	90%	70%	Helen Traubel (C)
			Enrico Caruso (V)
			Boston Pops Orchestra (V)

Comment: The noble and grandiose aria moves at a majestically measured pace from its first to its last tone.

TITLE AND COMPOSER	EMOTIONAL STRENGTH	LISTENER AGREEMENT	PERFORMER
4. Jesu, Joy of Man's Desiring—Bach	48%	50%	Stokowski with Orchestra (V)
			E. Power Biggs (organ) (C)

Comment: An early German chorale tune, a devotional song of simple and joyous faith, is the core of the music. The design of the flowing accompaniment, in triple meter, is derived from the chorale itself.

MOOD CATEGORY E (SAD, MELANCHOLY, GRIEVING, DEPRESSING, LONELY)

TITLE AND COMPOSER	EMOTIONAL STRENGTH	LISTENER AGREEMENT	PERFORMER
1. Sonata Opus 10, No. 3 (Second movement)—Beethoven	72%	64.3%	Kempff (piano) (D)

Comment: Largo é mesto—a sad, burdened movement, like a deep blue sea, with its hidden terrors, but also with the grandeur of its solitude.

TITLE AND COMPOSER	EMOTIONAL STRENGTH	LISTENER AGREEMENT	PERFORMER
2. Sonata Opus 35 (Funeral March)—Chopin	64%	60%	Casadesus (piano) (C) Horowitz (piano) (V) Rubinstein (piano) (V)

Comment: A solemn, resonant toll of a bell echoes beneath a tragic melody which moves in a dirge-like rhythm, scarcely leaving one tone. Accompanied by the persistent toll of the bell figure, the melody expands in sonority and volume as it rises in dissonant harmonies. Ascending chords struggle against the tragic motif, only to sink as the dirge is resumed. Suddenly, like a healing balm, there appears a lyric section whose transparent harmonies support a sweet melody. The somber theme ultimately returns, but its tragic insistence seems somehow less harsh in the light of the comforting measures that precede it.

TITLE AND COMPOSER	EMOTIONAL STRENGTH	LISTENER AGREEMENT	PERFORMER
3. Romeo and Juliet Overture—Tschaikowsky	74%	55.7%	Rodzinski, Cleveland Orchestra (C)

Stokowski, N.Y. Phil.-Symphony Orchestra (C)

Toscanini, N.B.C. Symphony Orchestra (V)

Comment: Somber, foreboding chords in the horns open the piece. They later expand into the strife theme, which overshadows the entire work. Soon a surging love theme is introduced in the strings and horns, accentuated by dissonances whose resolution is purposely delayed. The poignant quality of the melody is further intensified by a constant undercurrent of pathetic minor second intervals. As the tragedy unfolds, this theme evolves into the ultimate death struggle; it is heard as a dirge, and at last, after a comforting chorale, as a spiritualized reflection of itself.

TITLE AND COMPOSER	EMOTIONAL STRENGTH	LISTENER AGREEMENT	PERFORMER
4. Symphony No. 6 ("Pathétique") (Fourth movement)—Tschaikowsky	64%	54.5%	Rodzinski, N.Y. Phil.-Symphony Orchestra (C) Von Karajan, Vienna Philharmonic Orchestra (C) Toscanini, N.B.C. Symphony Orchestra (V)

Comment: An atmosphere of mourning and lamentation which in some places reaches the depth of utter despair, in others reflects a mood of resignation and apathy.

TITLE AND COMPOSER	EMOTIONAL STRENGTH	LISTENER AGREEMENT	PERFORMER
5. Symphony No. 3 in E flat major ("Eroica") (Second movement)—Beethoven	62%	53%	Schuricht, Berlin Philharmonic Orchestra (D) Walter, N.Y. Phil.-Symphony Orchestra (C)

MOOD CATEGORY E (SAD, MELANCHOLY, GRIEVING, DEPRESSING, LONELY) – (Continued)

Comment: A funeral march, one of the best known. Soft dialogue between the violin and the oboe. The music gradually gains in expression and ends in a paean of tremendous emotional intensity.

TITLE AND COMPOSER	EMOTIONAL STRENGTH	LISTENER AGREEMENT	PERFORMER
			Toscanini, N.B.C. Symphony Orchestra (V)
			Koussevitzky, Boston Symphony Orchestra (V)
6. Tristan und Isolde: Liebestod—Wagner	74%	51%	Stokowski with Orchestra (V)

Comment: A spirit of rapt reverie. Soft strings and horns create a pensive mood, to which the vibrant soprano voice brings deepened color. In a langorous rhythm, the voice and orchestra sing as one; the orchestra moves through ascending harmonies, rich in dissonance and emotional tone, which the voice spins out melodically. Together, with a yearning never resolved, they reach a peak of intensity. Then there is a gradual descent to resignation and peace.

TITLE AND COMPOSER	EMOTIONAL STRENGTH	LISTENER AGREEMENT	PERFORMER
			Rodzinski, Chicago Symphony Orchestra (V)
			Helen Traubel with Orchestra (V)
7. Symphony No. 7 in A (Second movement)—Beethoven	64%	50.9%	Toscanini, N.Y. Phil.-Symphony Orchestra (V)

Comment: Pensive sadness combined with restrained emotion. Calm melody evolves with glowing sincerity against a background of a distinct though simple rhythmic pattern.

MOOD CATEGORY F (EERIE, WEIRD, GROTESQUE)

TITLE AND COMPOSER	EMOTIONAL STRENGTH	LISTENER AGREEMENT	PERFORMER
1. Firebird Suite, Part I—Stravinsky	68%	60%	Stravinsky, N.Y. Phil.-Symphony Orchestra (C) Stokowski and Orchestra (V)

Comment: Mysterious whispers and murmurings indicate the enchanted orchard where the Fire Bird lives. The bird performs a dance to the accompaniment of glittering chords, until the entrance of the Prince causes the bird to withdraw.

TITLE AND COMPOSER	EMOTIONAL STRENGTH	LISTENER AGREEMENT	PERFORMER
2. La Mer (First movement)—Debussy	68%	59.5%	Ormandy, Philadelphia Orchestra (C) Toscanini, N.B.C. Symphony Orchestra (V)

Comment: A tone painting reflecting the awesome majesty and grandeur of the ocean. Spume-capped waves roar shoreward, carrying the briny scent and glittering spray. A bright expanse of blue stretches enigmatically to the vast horizons of man's desire.

TITLE AND COMPOSER	EMOTIONAL STRENGTH	LISTENER AGREEMENT	PERFORMER
3. The Rite of Spring (Part I)—Stravinsky	80%	56.3%	Stravinsky, N.Y. Phil.-Symphony Orchestra (C)

MOOD CATEGORY F (EERIE, WEIRD, GROTESQUE)—(*Continued*)

Comment: The plaintive voice of the bassoon opens the movement in which the mystery of the physical world is portrayed. The orchestra reproduces the mood of mystery, the stamping rhythms of the dance, which is accompanied by a persevering chord that sends chills down the listener's spine. The music rises to a furious climax, when a sudden hush of the orchestra indicates the entrance of the Sage who invokes the rejuvenation rite. Drums signal the Dance of Earth which grows more and more ecstatic.

TITLE AND COMPOSER	EMOTIONAL STRENGTH	LISTENER AGREEMENT	PERFORMER
4. Pierrot Lunaire—Schönberg	76%	52.8%	Stiedry-Wagner, soloist, with chamber orchestra conducted by Schönberg. (C)

Comment: Pale, ghost-like moon pictures, impalpable, like a phantom, dissolving in the night air.

ABOUT THE MUSIC RESEARCH
FOUNDATION

by Jay T. Wright, Ph.D.*

> Music is the essence of order, and leads to
> all that is good, just, and beautiful.—*Plato*

The Music Research Foundation, Incorporated, is a non-profit New York membership corporation which was founded in 1944. It is dedicated to the continuation and advancement of scientific research related to the use of music as an adjunct to medicine. The management of the Foundation is vested in a nine-member Board of Directors, among whom are leading figures in science, medicine, music, education, and business. The Research Committee is made up of distinguished physicians and psychiatrists, including such men as the Chairman of the Committee on Therapy of the American Psychiatric Association, the Director of the National Committee for Mental Health, and the Director of the Psychiatric Institute of New York.

The work of the Music Research Foundation began in April, 1944, when the Surgeon General of the Army authorized the group to test under medical

* Director of the Workshop in Intergroup Education, Columbia University Teachers College. Chairman, Board of Directors, Music Research Foundation.

supervision the effect of music on selected patients at Walter Reed General Hospital. For this purpose the hospital created a Department of Applied Music, which was directed by Miss Frances Paperte, founder of the Music Research Foundation, under the supervision of the Hospital Medical Staff.

The work at Walter Reed General Hospital continued for over three and a half years and resulted in such outstanding examples of favorable evidence of the helpfulness of music to the patients that there was no room for doubt that the therapeutic potentialities of music should be subjected to extensive scientific research.

The results obtained attracted widespread medical attention and brought requests for reliable knowledge from all over the world. In order to fulfill this demand, the Foundation decided not only to expand its own scope of activity, but also to coordinate and assist other organizations that might desire to do research in the field. Dr. R. C. Williams, Assistant Surgeon General, U. S. Public Health Service, assumed chairmanship of the Foundation's Board of Directors and of the Research Committee in 1947 to effect this expansion. Under Dr. Williams' leadership, the Music Research Foundation established a series of specific research projects designed to determine under laboratory conditions the value, use, and limitations of music as an aid to therapy.

While putting to immediate use the limited body

of knowledge that has come out of past research, the Foundation has from the outset concentrated its chief efforts in accumulating new data through basic research. By freely disseminating research results, the organization encourages others to apply specially selected music according to a predetermined plan based on the individual need.

Far from being discouraged by the inherent slowness in obtaining results in any field of research, the Foundation workers feel ever more confident that music can be shaped into a powerful new weapon for the social good, provided that premature exploitation of the field can be avoided and that applied music technicians will be patient until a sound technique can be evolved upon the secure basis of systematic scientific research.

NEED OF SCIENTIFIC RESEARCH

Although there has been some unrelated work in the use of music in occupational therapy and some experimentation with music in psychiatry, *the physician must have reliable data and proven techniques* if this valuable tool is to become a true adjunct to medicine.

In addition to its own specific research projects, the Music Research Foundation attempts to evaluate present knowledge, disseminate reliable information, and offer practical guidance and training in diversified phases of the problem.

THE FOUNDATION'S RESEARCH PROGRAM

The Research Program is operated through existing institutions of research and learning for research projects designated and defined or accepted and sponsored by the Foundation's Research Committee.

The current program of the Research Committee consists of efforts to

(1) Collect information and coordinate the efforts of those now engaged in activities related to the use of music as an adjunct to medicine;

(2) Encourage or develop further scientific research which will yield additional data in these and other promising areas;

(3) Apply music as an accessory therapeutic tool as extensively as possible on the basis of acquired reliable techniques;

(4) Train a nucleus of technicians in readiness for more extensive development of the field in the near and distant future.

THE CATTELL PROJECT

An interesting series of experiments, sponsored by the Music Research Foundation through the Helen Bonfils Fellowship, was conducted by Raymond B. Cattell and his associates at the University of Illinois. The investigation concerned the relation between musical preference and the listener's personality. Dr. Cattell proceeded from the supposition that an individual is influenced most effectively by a type of music which he considers most attractive.

In his report on these experiments, Dr. Cattell pointed out that the problem is one of classifying pieces of music according to their attractiveness to various kinds of patients and various kinds of personalities.

Whether or not it is correct to assume that the music likely to have the most desirable effects on a patient is that which he most desires, it is necessary to classify music for diagnostic purposes. Indeed, the object of this research is initially nothing more than diagnosis. When some firm knowledge is on hand regarding the use of music for diagnostic purposes, it will be appropriate to move on to the further question of the use of music for therapeutic purposes. Accordingly, the main object of this research is to demonstrate statistical relationships between musical preferences and types of listeners' personalities.

The present research unit has been working for ten to fifteen years on problems of personality diagnosis and has produced evidence, by means of factor analytic techniques, of some sixteen different dimensions of personality which can be measured either by questionnaires or by objective tests or by behavior ratings in everyday life situations. These sixteen factors, including schizothymia, general intelligence, surgency, dominance, etc., have been extensively described in many published researches. In the present research they were therefore the basic reference dimensions against which the musical preferences were compared. That is to say, every subject was measured on the sixteen personality factors by means of the questionnaire, and sometimes also by means of a humor test, and was assigned a score on each of the sixteen independent factors.

Now in order to relate musical preferences to the measured personality factors, it is necessary first to group musical preferences in such a way that independent dimensions of preference are first demonstrated. This can be done again by factor analysis, which employs the following procedure. Some sixty different pieces of music were played to one hundred persons, each person being asked to say whether he liked, disliked, or was indifferent in regard to each piece of music. Several thousands of correlation coefficients were then worked out representing every possible combination of the pieces of music in pairs. Thus we can always say to what extent liking for any one piece of music A is related to the liking for any other piece of music B. By inspecting the correlation matrix, one can find the clusters of pieces of music which "go together." Thus in one cluster there were some seventeen pieces of music, in which liking for one tended to go with liking for every one of the other sixteen pieces. By factor analysis it can be shown that such a cluster constitutes an independent dimension of musical preference, and any one individual can be scored on this dimension from zero to seventeen, according to whether he likes one or all

of the pieces concerned. About nine other such independent clusters were found, so that any individual
can receive a score on some nine distinct independent
dimensions of musical taste.

The final step in this research would be the correlation of the musical preference dimensions with
the personality dimensions. A part of the same analysis is a comparison of a group of normal subjects
with a group of one hundred subjects from mental
hospitals who were also put through the same test.
By this means Cattell hopes to discover whether the
profile on the nine measurements is significantly different for psychotics and normals and for particular
psychotic syndromes.

Although the final decision as to what dimensions
of musical taste exist must rest upon the statistical
analysis, Cattell and his associates were guided in the
initial stages of the experiment by classifications
which have already been made by musicians and
others. It was their aim to include in the original
sixty pieces of music a universal selection of exciting
music, soothing music, and so on. The work done by
Frances Paperte and others in the original Walter
Reed study was a useful guide in this preliminary
choice of musical selections, as also was the work
which had been done at Edinburgh by Dr. Vernon.
Because sixty selections would not give enough items
in the end for the scoring of as many as nine different
dimensions of taste, it was decided to add sixty more,
so that, in fact, the subjects listened to 120 pieces of

music. These were factorized in two groups, first a matrix of sixty and then a matrix of eighty, the overlap of twenty items being necessary to cement the second matrix to the first and to locate the factors found in one in relation to factors found in the other. This is only a statistical time-saving device in the interests of the final aim of discovering the factors, i.e., classifying the whole of the 120 pieces of music in their main categories or dimensions. Dr. Cattell plans to reproduce the pieces of music which are diagnostically most significant in a set of perhaps forty to seventy items on a slow playing record in order that this may be used as a diagnostic instrument by psychologists and psychiatrists.

THE COLUMBIA UNIVERSITY TEACHERS COLLEGE
PROJECT

A field study has been set up to investigate the influence of music upon individual and group behavior and the emotional adjustment of children. A program of musical activities and experiences is to be organized with a group of children, and analytical criteria are to be set up to ascertain the psychological effects of these activities and experiences. The project will be carried out in a nearby public school system and will be conducted during and throughout the school year 1952-1953. It is contemplated that the project will be continued throughout several subsequent years beyond the year 1952-1953.

The field work is to be conducted by two field

workers. One of them is an experienced public school music teacher, and the other is a qualified psychologist. The field workers will hold staff appointments at Teachers College and will be nominated and appointed by Teachers College after consultation with the Music Research Foundation. The responsibility of these workers will be (a) to conduct throughout the school year the program of musical activities and experiences already indicated, and (b) to apply suitable criteria to ascertain the psychological effects of these activities and experiences on the children. Teachers College is responsible for selecting the school system where the project is being conducted and has made suitable arrangements with the officials of the school system. An advisory committee jointly agreed upon, consisting of experts designated by the Music Research Foundation and of representatives of the departments of Music and Psychology of Teachers College, will have general supervision of the project and responsibility for shaping its policies, in particular for the supervision of the field workers and for selecting the analytical criteria to be employed in determining the psychological effects of the music program upon the children. Professor James L. Mursell will represent Teachers College and will be in administrative charge of the project. Professor Mursell will have the cooperation of Arthur T. Jersild, Professor of Education at the College and a committee member.

The project also proposes to study the value of

music as a medium for learning more about the growth process in children, to discover why many musically gifted boys and girls do not do well in other subjects, to help the child understand himself and to become well adjusted in school and at home, to learn why music is often an uplifting experience for some children and a dull and academic experience for others, and to discover how music can be made a more effective part of the school curriculum and of home and community life.

There is no truer truth obtainable
By Man, than comes of music.

—Browning

DR. GEORGE S. STEVENSON'S LETTER

THE NATIONAL ASSOCIATION FOR MENTAL HEALTH, Inc.

1790 Broadway, New York 19, N. Y.

President
OREN ROOT

Medical Director
GEORGE S. STEVENSON, M.D.

Treasurer
A. L. von AMERINGEN

March 27, 1952

Miss Frances Paperte
Music Research Foundation
2 East 63rd Street
New York 21, N. Y.

Dear Miss Paperte:

I am pleased to be able to make the following statement which the Music Research Foundation may use at its discretion.

The widespread occurrence of music among widely distributed peoples and varied cultures is evidence that in music we have a great psychological force. Most people recognize the deep effect of music upon human feeling.... *The question is, can music be harnessed so that its effects are more subject to our control?* Can it be made to do our bidding, to give poise to the tense soul, to raise one from depression, or to soften excitement? Are these results inherent in music or in the make-up and experience of the

97

person hearing it? In order to have a desired effect, must the music be chosen in keeping with the idiosyncrasies of the listener? These are questions facing psychiatry. But they are also a challenge facing the joint effort of music and science to solve them. These questions are the task of the Music Research Foundation now joining in a project with Columbia University looking toward their solution. The ways of research are usually rough, and one can anticipate a long-time effort to find answers to these questions. The answers will come bit by bit with the pursuit of many false leads in between, but the effort is worth while, for out of it we may expect light to be thrown on a large and important aspect of human life.

Sincerely yours,

(Signed)   George S. Stevenson, M.D.
            Medical Director.

APPENDIX A

The MELO Quadricircle

by Emil A. Gutheil, M.D.

The following is a classification of some of the possible variables in the study of the effects of music on behavior. It covers four principal areas of research: M, materials; E, effects; L, listener; O, objectives; and it is expected to provide a framework within which organized research may be undertaken.

M—MATERIALS

A. Setting

    I. Place (institute)......

    II. Time:

        1. Date......

        2. Hour of day......

    III. Participants in experiment:

        1. Investigator(s):

            a. Chief......

            b. Worker......

            c. Supervision......

            d. Controls......

        2. Subject (s):

            a. Number......

            b. Sex......

            c. Remarks......

IV. Room:

    1. Soundproofing:
        a. Present . . . . . .
        b. Absent . . . . . .
        c. Remarks . . . . . .

    2. Acoustics of room:
        a. Reverberation time . . . . . .
        b. Sound absorption:
            (1) Loss in db . . . . . .
            (2) Absorption in per cent . . . . . . in
                cycles . . . . . .
            (3) Other factors . . . . . .
        c. Floor space in sq. ft. . . . . . .
        d. Other details . . . . .

B. Source:

    I. Single sound . . . . . .

    II. Composition (selection):

        1. Name of composer . . . . . .
        2. Name of selection . . . . . .

    III. Performance:

        1. Direct ("live") . . . . . .
        2. Indirect (reproduced) . . . . . .
        3. Name of main performer . . . . . .
        4. Name of band (orchestra, choir) . . . . . .
        5. Remarks . . . . . .

    IV. Aspect:

        1. Qualitative:
            a. Vocal . . . . . .
                (1) Voice(s) . . . . . .
                (2) Words

APPENDIX A

    (a) Native, foreign [Underscore]
    (b) Not used......
  b. Instrumental......
2. Quantitative:
  a. Solo......
  b. Ensemble (also Solo + Accomp.)
    (1) Small group (chamber music, other)......
    (2) Large group (choir, band, symphony, other)......
3. Historical:
  a. Composition contemporary......
  b. Century......
  c. Remarks......

C. Vertical structure of selection ("Space Factor"):
  I. Timbre (tone quality, overtones)......

  II. Volume (amplitude, intensity, loudness) [in db]......
  1. Total volume evaluated......
  2. Circumscribed area evaluated......
    a. Phrasing:
      (1) Composer's indication......
      (2) Performer's:
        (a) Portamento vibrato......
        (b) Rate in vibrato cycles......
        (c) Remarks......
    b. Other factors......
  3. Character of volume......
  4. Sound level:
    a. Ambient noise level......
    b. Music intensity level......
    c. Iso-Reftone level......

III. Harmony (Chords):
    1. Sonority (sonance):
        a. Consonance......
        b. Dissonance......
        c. Mixed......
        d. Remarks......

    2. Character......
    3. Harmonic interval......

IV. Pitch (frequency) [in cycles per second]
    1. High, low, medium [Underscore]
    2. Pitch vibrato......

V. Register

D. Horizontal structure of selection ("Time Factor")
  I. Rhythm:
    1. Quantitative aspect:
        a. Beat......
        b. Meter time (measure)......
        c. Phrase......
        d. Remarks......

    2. Qualitative aspects:
        a. Constant......
            (1) Simple......
                (a) Dance......
                (b) March......
                (c) Other......
            (2) Syncopated......
            (3) Other......
        b. Variable......

    3. Accents:
        a. Type (metric, dynamic) [Underscore]

APPENDIX A

    b. Position (balanced, off-balance) [Under-score]

    c. Remarks......

II. Tempo (metronome reading)......

    1. Specified (Largo, Adagio, Andante, Moderato, Allegro, Presto, Other)......

    2. Unspecified (generally slow, generally fast, medium fast) [Underscore]

    3. Variable......

    4. Remarks......

III. Melody:

    1. Movement:

        a. Rising......

        b. Falling......

        c. Mixed......

        d. Counterpoint......

    2. Style:

        a. Distinct......

            (1) Liturgic......

            (2) Classical......

            (3) Romantic......

            (4) Popular......

            (5) Other......

        b. Indistinct......

    3. Mode:

        a. Major......

        b. Minor......

        c. Mixed......

        d. Remarks......

    4. Ending......

IV. Length of the selection performed:

    1. Complete......

2. Incomplete (tone, chord, melodic interval, figure, period) [Underscore]
3. Remarks......

V. Selection used:
1. One......
2. More [state number]......

VI. Duration of performance (timing)......

E. General character of selection

I. Form (Etude, Sonata, Lied, Scale, Other)......

II. Mood [gay (figures of joy), sad (figures of lamentation), indifferent, other]......

III. Type of selection ("absolute," "program," tonal, atonal, entertainment, other)......

F. Music used:

I. Alone......

II. In conjunction with other procedures......

1. Therapy:
a. Psychotherapy......
(1) Interview type......
(2) Hypnosis......
(3) Other forms......
b. Surgery (incl. dentistry)......
c. Hydrotherapy......
d. Occupational therapy......
e. Physiotherapy......
f. Rehabilitation......
g. Other procedures......
2. Teaching situations......

APPENDIX A

    3. Dance......
    4. Other......

G. Controls:

    I. Noise......

    II. Other stimuli:
        1. Speech......
        2. Sound......
        3. Other......

    III. Statistical controls

E—EFFECTS

Before After

A. Duration of stimulus......

B. Psychologic effects......

    I. Behavior.......
        1. Group......
           a. Descript......
           b. Non-descript......
        2. Individual......
           a. Subjective effects......
           b. Reaction toward others......
           c. Other effects......

    II. Mood:
        1. Sedative:
           a. Relaxing......
           b. Soporific......
           c. Depressing......
           d. Other......
        2. Stimulating:
           a. Tension increasing......

Before   After

b. Invigorating......
c. Irritating......
d. Uplifting......
e. Other.......

3. Indifferent......

4. Remarks......

III. Reaction Time......

IV. Effect on suggestibility......

V. Effect on distractibility......

VI. Output of activity......

1. Vocal (talking, shouting, singing, others)......

2. Motor......
a. Entire body......
b. Parts of body......
c. Remarks......

VII. Social attitude......

VIII. Educational effects......

1. Scholastic achievement......
2. Integrative ability......
3. Growth (maturation)......
4. Memory......
5. Perception and apperception......
6. Imagination......
7. Attention......
8. Discipline......
9. Habits......

APPENDIX A

Before After

    10. Reasoning. . . . . .
    11. Language. . . . . .
    12. Special abilities. . . . . .
    13. Spiritual uplifting. . . . . .
    14. Other. . . . . .

  IX. Other effects. . . . . .

C. Physiologic effects. . . . . .
  I. Systemic. . . . . .
    1. Organ group:
      a. Sensorium. . . . . .
      b. Cardiovascular. . . . . .
        (1) Blood pressure. . . . . .
        (2) Pulse rate. . . . . .
        (3) Amplitude of pulse. . . . . .
        (4) Remarks. . . . . .
      c. Respiratory. . . . . .
        (1) Rate. . . . . .
        (2) Amplitude. . . . . .
        (3) Remarks. . . . . .
      d. Gastro-intestinal. . . . . .
      e. Nervous system. . . . . .
        (1) Central. . . . . .
        (2) Peripheral. . . . . .
        (3) Vegetative. . . . . .
        (4) Pain response. . . . . .
        (5) Temperature
            response. . . . . .
        (6) Remarks. . . . . .
      f. Skeletal muscle. . . . . .
        (1) Muscle tone. . . . . .
        (2) Muscle spasm. . . . . .
        (3) Contracture. . . . . .

Before   After

    (4) Movements......
        (a) Voluntary......
        (b) Involuntary......
        (c) Remarks......
    (5) Endurance......
    (6) Fatigue......
    (7) Reflexes......
        (a) Superficial......
        (b) Deep......
        (c) Remarks......
    (8) Accuracy......
    (9) Speed (psychomotor)......
    (10) Tremor (type)......
    (11) Remarks......
  g. Smooth muscle......
    (1) Direct influence......
    (2) Conditioned reflex......
  h. Skin......

2. Metabolism
  a. Blood constituents......
    (1) Glucose and glucose
        tolerance......
    (2) CO_2 and O_2......
    (3) Leucocytic response......
    (4) Hematocytic response......
    (5) Remarks......
  b. BMR......
  c. Glandular secretions......
  d. Remarks......

3. Speech

II. Synesthetic effects:

  1. Chromesthesia......

APPENDIX A

    2. Cutaneous reflex......

    3. Other......

  III. Other physiologic effects......

D. Threshold for stimulus......

  I. Physiologic......

  II. Psychologic......

E. Specific effects......

  I. Qualitative......

    1. Visual......

    2. Auditory......

    3. Ideational......

    4. Kinesthetic......

    5. Other......

  II. Quantitative (vividness)......

F. Stress conditions......

  I. Adjustment to stress......

  II. Frustration effects......

G. Response to therapy......

L—LISTENER

A. Personality type assessed:

  I. Cultural......

  II. Physical:

    1. Kretschmer......

  2. Anthropologic (racial)......
  3. Social.....
  4. Sheldon......
  5. Other......

 III. Psychologic:

   1. Jung......
   2. Other......

B. IQ......

  I. Individual......

  II. Group......

  III. Other......

C. Personal data......

  I. Vocation.....;
  II. Aptitude......
  III. Interests......
  IV. Other data......

D. State of mind

  I. "Normal"......

  II. "Abnormal"......

   1. Psychic......
    a. Neurosis......
    b. Psychosis......
    c. Mental deficiency......
    d. Psychopathy......
    e. Other......

   2. Organic:
    a. Exclusively organic......

    b. Combined with psychic......

    c. Remarks......

E. Mood:

    I. Predominantly stable......

    II. Predominantly variable......

F. Physical condition:

    I. Normal......

    II. Abnormal......

        1. Posture......

        2. Locomotion......

    III. Physical status......

        1. General......

        2. Auditory......

        3. Neurologic......

        4. Fatiguability......

        5. Other factors......

    IV. Other findings......

G. Receptivity to music:

    I. Present......

        1. Listener practicing......

            a. Professional......

            b. Amateur......

            c. Musical training......

            d. Musical aptitude......

        2. Non-practicing

    II. Absent......

    III. Remarks......

MUSIC AND YOUR EMOTIONS

H. Familiarity with selection:
   I. Familiar......
      1. Can identify selection......
      2. Cannot identify selection......
      3. Can identify instrument(s) (voice)......
      4. Cannot identify instrument(s) (voice)......

   II. Partly familiar......

   III. Unfamiliar......

I. Participation:
   I. Active (subject takes part in performance)......
      1. Degree of participation......
      2. Other details......

   II. Passive......
      1. Attention......
      2. Distractibility......
      3. Associations......
      4. Daydreams......
      5. Memory for selection......

J. Attitude toward selection:
   I. Quality......

   II. Wish to hear it......

   III. Reason for (II)......

   IV. Pleasure-Displeasure......
      1. Selection agreeable ......
      2. Selection disagreeable......
      3. Indifferent......
      4. Remarks......

APPENDIX A

K. The subject:
    I. Age. . . . . .

    II. Sex. . . . . .

    III. Race. . . . . .

    IV. Nationality. . . . . .

    V. Educational level. . . . . .

    VI. Religion. . . . . .

    VII. Other data. . . . . .

L. Frequency of hearing:
    I. First. . . . . .

    II. Repeated. . . . . .
        1. After relat. short interval. . . . . .
        2. After relat. long interval. . . . . .

M. Identifying of composer's emotion:
    I. Possible. . . . . .

    II. Partly possible. . . . . .

    III. Impossible. . . . . .

N. Attitude toward performer. . . . . .

O. Attitude toward experiment. . . . . .

P. Subject's explanation of effect. . . . . .

MUSIC AND YOUR EMOTIONS

I. Effect attributed to:

    1. Total selection......
    2. Parts of selection......
       a. Rhythm......
       b. Harmony......
       c. Melody......
       d. Other factors......

    3. Performer's personality......
    4. Remarks......

II. Unable to determine......

Q. Heredity:

I. Father:

    1. Musical......
       a. Practicing......
       b. Non-practicing......
       c. Remarks......

    2. Non-musical......
    3. Remarks on father......

II. Mother:

    1. Musical......
       a. Practicing......
       b. Non-practicing......

    2. Non-musical......

    3. Remarks on mother......

III. Has the family as a group expressed a definite attitude toward music?......

IV. Other details......

APPENDIX A

R. Type of judgment:
   I. Objective [Example: "Perfect blending"] . . . . .
   II. Subjective ["It makes me sad"] . . . . . .
   III. Associative ["It resembles church bells"] . . . . . .
   IV. Other . . . . . .

O—OBJECTIVES

A. Measuring instruments used:
   I. Medical:
      1. Physical [Underscore]: Algesiometer, audiometer, BMR, blood chemistry, capillograph, ECG, EEG, EMG, ergometer, fluoroscope, gastroscope, oscillometer, plethysmograph, psychogalvanic reflex measurement, pupillograph (Lowenstein), respirograph (acoustic) [Kubie-Margolin], RR, skin temperature measurement, stomach acidity test, thermometer, X-ray, Other . . . . . .
      2. Psychologic [Underscore]: Rorschach charts, TAT charts, stopwatch, performance test, tachistoscope, Other . . . . . .
   II. Musical [Underscore]: Amplifier, harmonic analyzer, metronome, microphone, oscillograph, output (intensity) meter, phonophoto camera, piano camera, rhythm meter, recording machine, sound spectrograph (Potter, Kopp, Green), tone generator, tonoscope, tuning fork, Other . . . . . .
   III. Other instruments . . . . . .

B. Research:

    I. Tabulating information regarding previous research......

    II. Allocation and organization of personnel......

    III. Establishment of terminology:
        1. Classification of selections......
        2. Classification of subjects......
        3. Other classifications......

    IV. Coordinating existing measurements regarding:
        1. Materials:
            a. Musicology......
            b. Acoustics......
            c. Psychology......
            d. Education......
            e. Other......

        2. Subjects:
            a. Psychologic tests......
            b. Interview......
            c. Questionnaire......
            d. Physiologic tests......
            e. Other......

        3. Effects:
            a. Meaning to listener......
            b. Response by listener......
            c. Mechanism of transformation of tone perception into psychologic and psychosomatic reactions......

        4. Statistics......

5. Laboratory experiments......
6. Predictions......
   a. Explanatory prediction......
   b. Trends prediction......

V. Educational evaluation......

VI. Social evaluation:
   1. Tests......
   2. Studies in cultural setting......
   3. Records......
      a. Physical......
      b. Social......
      c. Personal......
   4. Preparation of case studies......
   5. Evaluation of methods......
   6. Other factors......

VII. Coordinating criteria of interpretation......

VIII. Criteria for keeping records......

IX. Establishing policy regarding:
   1. Publication of results......
   2. Exchange of information......

X. Determining research projects......

XI. Allocation:
   1. Of materials (subjects)......
   2. Of funds......
   3. Other allocations......

XII. Classification of individual responses:
   1. Objective......
   2. Associative......

      3. Imaginative......

      4. Sensorial......

      5. Other......

XIII. Graphic representation of reactions:

      1. Pitch.... .

      2. Volume......

      3. Duration......

      4. Tonal spectra (timbre)......

      5. Psychologic......

      6. Physiologic......

      7. Other graphs......

XIV. Establishment of training standards (curricula, etc.) for music workers......

XV. Establishing a policy regarding use of grants and fellowships......

XVI. Other issues......

C. Diagnosis:

    I. Musical personality tests......

    II. Other......

D. Therapy:

    I. Formulation of stimulus-reaction laws......

    II. Practical tests of theory pertaining to cause and effect......

    III. Development and test of the procedure to insure predictability......

APPENDIX A

IV. Applications:
1. Psychotherapy......
2. Pain control......
3. Occupational therapy......
4. Industry......
5. Organic medicine......
6. Preventive disease......
7. Military medicine......
8. Other objectives......

V. Organization of therapeutic services:
1. Medical......
2. Psychologic......
3. Nursing......
4. Musical......
5. Administrative......
6. Other......

VI. Structuring of effective therapeutic program patterns

VII. Scheduling of therapeutic sessions

VIII. Personal qualifications of the music worker:
1. Musical background......
2. Personality evaluation......
3. Attitude for work with subjects......
4. Attitude toward subjects......
5. Functioning in therapeutic situation......
6. Experience in therapy......
7. Other......

APPENDIX B

The following list indicates the reaction of the students to the 105 musical selections used by Dr. Capurso in his study. Beneath the title of the selection in question is listed the number and manufacturer of the record actually used by Dr. Capurso.

Opposite the name of the piece of music we have listed the percentage of subjects placing the selection in each of the five mood categories. The *superscript* number indicates the number of musical experts who chose that category for the selection.

The mood categories are as follows:

- A: Happy, Gay, Joyous, Stimulating, Triumphant
- B: Agitated, Restless, Irritating
- C: Nostalgic, Sentimental, Soothing, Meditative, Relaxing
- D: Prayerful, Reverent
- E: Sad, Melancholy, Grieving, Depressing, Lonely
- F: Eerie, Weird, Grotesque

TITLE, COMPOSER, AND RECORD NUMBER—MOOD CATEGORIES

Stars and Stripes Forever (Sousa)—Decca 2132-A—A, 93.0^6; B, 7.0.

William Tell: Overture (Finale) (Rossini)—Victor 20607-B— A, 91.3; B, 6.54^4; F, 2.2.

APPENDIX B

Symphony No. 5, 2nd mvt. (Shostakovitch)—Victor 16631-B—A, 88.7[3]; B, 9.0[7]; F, 2.2.

Symphony No. 35, 4th mvt. (Mozart)—Columbia MM-478-1—A, 88.6[3]; B, 11.4.

Washington Post March (Sousa)—Decca 2133-B—A, 88.2[4]; B, 6.7; C, 1.7; E, 1.7; F, 1.7.

Aida: Grand March (Verdi)—Victor 11885-A—A, 87.3[7]; B, 3.1; C, 4.0; D, 2.7; E, 1.4; F, 1.5.

Lohengrin: Prelude to Act III (Wagner)—Victor 7386-B—A, 87.2[6]; B, 10.7; F, 2.1.

Symphony No. 7, 3rd mvt. (Beethoven)—Columbia MM-577-7—A, 86.0[3]; B, 8.0; C, 6.0.

Moto Perpetuo (Paganini)—Victor 14325-B—A, 81.9[5]; B, 13.6; C, 4.5.

Piano Concerto No. 5, 3rd mvt. (Beethoven)—Victor 11-8322-B—A, 79.1[4]; B, 12.5; C, 8.4.

Incidental Music to A Midsummer Night's Dream: Scherzo (Mendelssohn)—Columbia MM-504-4, 11789-D—A, 77.8[6]; B, 14.5; C, 0.0[2]; F, 7.7[2].

Symphony No. 5, 4th mvt. (Beethoven)—Victor 8511-B—A, 76.7[14]; B, 17.8[2]; C, 2.4; D, 0.5; E, 1.3; F, 1.3.

Polonaise in A flat major, Opus 53 (Chopin)—Victor 11-8848-A—A, 73.5[3]; B, 12.2[3]; C, 10.2; E, 4.1.

Symphony No. 41, 4th mvt. (Mozart)—Columbia MM-565-6—A, 73.1[6]; B, 17.1; C, 9.8.

Humoresque (Dvorak)—Victor 1170-B—A, 73.0[4]; B, 15.9; C, 7.9; E, 1.6; F, 1.6.

Symphony No. 5, 1st mvt. (Beethoven)—Victor 15831—A, 70.4[8]; B, 9.1[2]; 15.9; E, 2.3; F, 2.3.

Symphony No. 4, 1st mvt. (Brahms)—Columbia MM-335-1—A, 69.6; B, 19.6; C, 10.8; E, 0.0[3].

Turkey in the Straw (Folk Tune)—Victor 4390—A, 69.1[2]; B, 0.0[1]; C, 27.6; F, 3.3.

Hallelujah Chorus, Messiah (Handel)—Victor 35767-B—A, 68.7[8]; B, 2.1; C, 2.1; D, 25.0; E, 2.1.

Etude in G flat major, Opus 10, No. 5 (Chopin)—Columbia M-368—A, 65.5[3]; B, 29.4; C, 3.4; E, 1.7.

Ritual Dance of Fire (DeFalla)—Victor 10-1135-A—A, 63.9; B, 27.9[2]; F, 8.2.

An American in Paris (Gershwin)—Victor 35963-A—A, 61.4[4]; B, 28.1; C, 7.0; E, 1.8; F, 1.8.

Mefisto Waltz (Liszt)—Victor 18409-A—A, 60.0; B, 27.7; C, 3.1; D, 1.5; F, 7.7[3].

Polka from "The Age of Gold" (Shostakovitch)—Victor 11-8239-B—A, 60.0; B, 15.2[8]; C, 1.4; D, 0.3; E, 0.7; F, 22.4[4].

Blue Danube Waltz (Johann Strauss)—Victor 16658-A, 16659-A—A, 56.9[2]; C, 42.5; F, 0.6.

Golliwog's Cake Walk (Debussy)—Victrola 7148-A—A, 54.5; B, 27.3; C, 5.5; E, 3.6; F, 9.1[2].

Fantaisie-Impromptu, Opus 66 (Chopin)—Victor 10-1141-A—A, 53.1[3]; B, 21.2; C, 19.7; E, 6.0.

Ride of the Valkyries, from Die Walküre (Wagner)—Victor 9163-A—A, 50.0; B, 32.2[2]; E, 2.9; F, 14.9.

Saber Dance from Gayne Suite No. 1 (Khachaturian)—Columbia 12498-D—A, 48.8[3]; B, 31.7[4]; C, 5.0; F, 15.0[2].

Etude, Op. 10, No. 12 (Chopin)—Columbia M-368—A, 48.0; B, 22.0[4]; C, 28.0; E, 2.0.

Till Eulenspiegel (Richard Strauss)—Victor 11724-A—A, 45.6; B, 32.6[2]; C, 2.2; F, 19.6[3].

Symphony No. 3, 1st mvt. (Beethoven)—Columbia MM-449-1 —A, 45.4[3]; B, 25.4; C, 10.9; D, 1.9; E, 14.5; F, 1.9.

Hungarian Rhapsody No. 2 (Liszt)—Victor 14422-A—A, 43.5[3]; B, 15.0[3]; C, 17.4[2]; E, 23.9.

APPENDIX B

Bolero (Ravel)—Columbia MX-22-1—A, 35.3[2]; B, 28.6[13]; C, 5.1; D, 0.8; E, 5.4; F, 24.8[2].

Hall of the Mountain King, Peer Gynt Suite No. 1 (Grieg)—Columbia MX-180-4—A, 33.3[2]; B, 35.3; C, 2.0; E, 5.9; F, 23.5[3].

Closing Scene, Götterdämmerung (Wagner)—Victor 6625-B—A, 28.3; B, 16.7; C, 21.8; D, 15.0[4]; E, 13.3; F, 5.0.

Finale, Symphonie Fantastique (Berlioz)—Victor 11-9033-A—A, 30.2; B, 58.1; C, 4.7; D, 2.3; F, 4.7[4].

Flight of the Bumble-Bee (Rimsky-Korsakoff)—Victor 11-9009-B—A, 42.6; B, 55.3[3]; E, 2.1.

Baba-Yaga (Liadow)—Brunswick 90049—A, 19.2; B, 57.4; E, 2.1; F, 21.3[3].

Sonata, Op. 35, 1st mvt. (Chopin)—HMV DB-2019—A, 13.0; B, 53.7[3]; C, 16.6; E, 13.0; F, 3.7.

Scheherazade, 4th mvt. (Rimsky-Korsakoff)—Columbia MM-398-1—A, 21.8; B, 49.1[2]; C, 3.6; E, 3.7; F, 21.8.

Pictures at an Exhibition, No. 10 (Moussorgsky)—Victor 7374-B, M-102—A, 27.0[1]; B, 47.6; C, 0.0[2]; D, 1.6; F, 23.8.

Der Erlkönig (Schubert)—Victor 15825-A—A, 7.1; B, 42.8[3]; C, 2.4; E, 40.5; F, 7.2[2].

Symphony No. 2, 2nd mvt. (Brahms)—Victor 11-9244-A—B, 2.2; C, 43.2[12]; D, 18.2; E, 36.4.

Sonata, Op. 110, 3rd mvt. (Beethoven)—HMV DB-7368—B, 11.4; C, 47.8; D, 4.5; E, 31.8[3]; F, 4.5.

Pavane pour une enfante défunte (Ravel)—Columbia 7361-M—A, 2.0; C, 68.0; E, 30.0[5].

The Swan, from Carnival of the Animals (Saint-Saëns)—Victor 1143-A—A, 1.5; C, 66.7[5]; D, 3.0; E, 27.3; F, 1.5.

Rhapsody in Blue—Ballad Theme (Gershwin)—Victor 13835-B—A, 6.4; B, 2.1; C, 78.8[11]; D, 2.1; E, 10.7.

Liebestraum No. 3 (Liszt)—Victor 11-8851-b—A, 2.1; B, 6.4; C, 76.6[5]; D, 4.3; E, 10.6.

Serenade (Schubert)—Columbia 7183-m—A, 0.0[1]; C, 70.6[2]; D, 5.9; E, 23.5.

Clair de Lune (Debussy)—Victor 11-8851—A, 4.6; B, 3.0; C, 69.4[35]; D, 3.1; E, 19.1[2]; F, 0.8.

Lullaby (Brahms)—Victor 20174-a—B, 5.8; C, 69.2[10]; D, 15.4; E, 9.6.

Air for the G String (Bach)—Columbia MM-703-6—A, 5.9; C, 56.9[6]; D, 19.6; E, 17.6.

Lost Weekend Score (Rozsa)—Victor 46-0000-a—A, 15.2; B, 8.5; C, 54.2; D, 1.7; E, 15.3; F, 5.1[3].

Moonlight Sonata, 1st mvt. (Beethoven)—Columbia MX-273-1—A, 1.8; B, 3.4; C, 54.1[21]; D, 8.5; E, 31.2; F, 1.0.

Symphony No. 8, 1st mvt. (Schubert)—Victor 6663-a—A, 14.3[2]; B, 7.1[2]; C, 52.4[11]; D, 2.4; E, 11.9; F, 11.9.

Scheherazade Suite, 1st mvt. (Rimsky-Korsakoff)—Columbia MM-398-1—A, 8.0; B, 6.7[3]; C, 52.0; E, 26.6; F, 6.7.

Symphony No. 5, 2nd mvt. (Beethoven)—Victor 9453-a—A, 25.5; B, 4.5; C, 48.7[10]; D, 5.6; E, 15.2; F, 0.5.

Valse Triste (Sibelius)—Victor 6579-a—A, 18.8; B, 3.4; C, 44.9; D, 1.8; E, 27.4[8]; F, 3.7.

Afternoon of a Faun (Debussy)—Victor 17700-a—A, 1.8; B, 3.7; C, 42.6[11]; D, 1.9; E, 38.9[2]; F, 11.1.

Asa's Death from Peer Gynt Suite No. 1 (Grieg)—Columbia MX-180-2—A, 1.9; C, 38.5; D, 23.1; E, 34.6[3]; F, 1.9.

Panis Angelicus (Franck)—HMV DB-1095—B, 4.7; C, 38.1; D, 28.6[8]; E, 28.6.

Symphony No. 9, 4th mvt. (Beethoven)—Victor 12-0062-b--—A, 25.9[11]; B, 9.7; C, 37.6; D, 13.0; E, 12.1; F, 1.7.

Daybreak from Daphnis and Chloe Suite No. 2 (Ravel)—Victor 18456-a—A, 6.8; B, 37.3[4]; C, 15.2; E, 3.4; F, 37.3.

APPENDIX B

Piano Concerto, 2nd mvt. (Grieg)—Columbia MM-313-4—A, 1.5; B, 3.1; C, 36.9⁴; D, 21.6; E, 36.9.

Symphony No. 6, 2nd mvt. (Beethoven)—Victor 11-9012-A—A, 17.3²; B, 15.4; C, 50.0⁴; D, 1.9; E, 15.4.

Symphony No. 5, 2nd mvt. (Tschaikowsky)—Columbia M-406-4, 5, 6—B, 4.0; C, 46.7¹³; D, 8.0; E, 38.6³; F, 2.7.

Waltz of the Flowers, from The Nutcracker Suite (Tschaikowsky)—Columbia MM-395, 70069—A, 2.5; C, 58.9³; E, 38.6.

Organ Choral No. 1 (Franck)—(English) Columbia CLX-2002—B, 2.1; C, 10.7; D, 78.7²; E, 8.5.

Mass in B Minor, Crucifixus (Bach)—Victrola 9966-A—C, 6.7; D, 71.1⁶; E, 20.0; F, 2.2.

Xerxes: Largo (Handel)—Victor 6648-A—A, 4.0; C, 14.0; D, 70.0⁴; E, 12.0.

Jesu, Joy of Man's Desiring (Bach)—Victor 4286-A—A, 25.0; B, 3.9; C, 15.4; D, 50.0⁴; E, 3.8; F, 1.9.

Hansel and Gretel: Children's Prayer (Humperdinck)—Victor 22176-A—A, 1.4; B, 2.9; C, 31.9; D, 43.5⁵; E, 20.3.

Ave Maria (Schubert)—Victor 11-9571-A—A, 2.9; B, 1.1; C, 33.0; D, 41.7¹³; E, 21.0; F, 0.3.

Requiem Mass: Requiem and Kyrie (Verdi)—Victor 17580-A—A, 2.9³; C, 10.0; D, 34.3; E, 37.1; F, 15.7.

Sonata, Op. 10, No. 3, 2nd mvt. (Beethoven)—HMV DB-8382—B, 9.5; C, 19.0; D, 2.4; E, 64.3³; F, 4.8.

Sonata, Op. 35, Funeral March (Chopin)—HMV DB-2020—A, 4.4; B, 8.9; D, 6.7; E, 60.0⁶; F, 20.0.

Romeo and Juliet Overture (Tschaikowsky)—Columbia MM-478-1—C, 15.4⁹; D, 23.1; E, 55.7; F, 5.8.

Symphony No. 6, 4th mvt. (Tschaikowsky)—Victor 16749-B—A, 3.0; B, 7.0; C, 26.2; D, 6.3; E, 54.5¹⁶; F, 3.0.

Symphony No. 3, 2nd mvt. (Beethoven)—Victrola 9453-A—B, 7.8; C, 21.6; D, 3.9; E, 53.0^4; F, 13.7.

Tristan und Isolde: Liebestod (Wagner)—Victor 8859-A—B, 4.1; C, 16.3^4; D, 28.6; E, 51.0^3; F, 0.0^5.

Symphony No. 7, 2nd mvt. (Beethoven)—Columbia MM-557-4—A, 3.4; B, 3.4; C, 16.9^3; D, 16.9; E, 50.9^3; F, 8.5.

Komm, süsser Tod (Bach)—Victor 16631-B—C, 26.7; D, 24.4^4; E, 48.9^2.

Finlandia (Sibelius)—Victrola 9015-A—B, 19.5; C, 4.9; D, 0.0^4; E, 48.8^4; F, 26.8.

Parsifal: Good Friday Music (Wagner)—Victrola 7163-B—A, 1.0; B, 7.8; C, 32.7; D, 8.6; E, 48.2^6; F, 1.7.

Le Gibet, from Gaspard de la Nuit (Ravel)—Columbia 69659-D—B, 9.1; C, 27.2^3; D, 6.1; E, 48.5; F, 9.1.

Isle of the Dead, Pt. I (Rachmaninoff)—Victor 7219-A—B, 6.8; C, 6.8^2; D, 4.5; E, 47.8^4; F, 34.1.

Saraband from English Suite No. 3 (Bach)—Victor 16630-B—B, 3.1; C, 32.8; D, 11.0^2; E, 45.3; F, 7.8.

None But the Lonely Heart (Tschaikowsky)—HMV E-534—A, 2.3; B, 6.6; C, 38.6^2; D, 6.6; E, 44.9^{14}; F, 1.0.

Symphony in D minor, 1st mvt. (Franck)—Columbia MM-608-1—A, 3.6; B, 8.9; C, 21.4^7; D, 14.3; E, 46.4; F, 5.4.

Lullaby from Gayne Suite No. 1 (Khachaturian)—Columbia 12500-D—B, 9.8; C, 31.4^7; D, 2.0; E, 43.1^3; F, 13.7.

Ma Mère L'Oye Suite, Part I (Ravel)—Victor 7370, 7371—B, 5.6; C, 25.9; D, 5.6; E, 40.7^2; F, 22.2.

Symphony No. 5, 2nd mvt. (Dvorak)—Victor 8738-A—A, 9.9; B, 3.7; C, 38.4^4; D, 6.7^8; E, 40.1; F, 1.2.

Symphony No. 1, 4th mvt. (Brahms)—Columbia MM-621-7—A, 6.0; B, 30.0; C, 18.0^2; D, 4.0^2; E, 40.0; F, 2.0.

Mathis der Mahler, 1st mvt. (Hindemith)—Victor 18333-A—A, 2.3; B, 15.9^2; C, 15.9; D, 2.3; E, 34.2; F, 29.4^4.

APPENDIX B

Russian Easter Overture (Rimsky-Korsakoff)—Victor 7018-A —A, 20.0; B, 11.1^2; C, 24.5; D, 4.4; E, 31.1; F, 8.9.

Firebird Suite, Part I (Stravinsky)—Victor 8926-A—A, 3.0; B, 10.6^2; C, 13.7; E, 12.1; F, 60.0.

La Mer, 1st mvt. (Debussy)—Victor 11650-A—A, 2.3; B, 1.5; C, 32.1^{10}; E, 4.6; F, 59.5.

Danse Macabre (Saint-Saëns)—Columbia 11251-D—A, 32.4; B, 21.8^3; C, 5.1; D, 0.8; E, 9.6; F, 30.7^{17}.

Rite of Spring, Part I (Stravinsky)—Columbia M-417-2—A, 3.4; B, 31.6^{12}; C, 1.8; D, 0.6; E, 6.7; F, 56.3^6.

Pierrot Lunaire (Schönberg)—Columbia MM-461-1—B, 41.5^2; E, 5.7; F, 52.8.

Soviet Iron Foundry (Mossolov)—Victor 4378-B—A, 4.3; B, 42.4; C, 1.0; D, 0.6; E, 4.6; F, 47.1^2.

The Sorcerer's Apprentice (Dukas)—Victor 7021-A—A, 4.0; B, 20.0; C, 2.2; E, 26.7; F, 46.7^5.

A Night on Bald Mountain (Moussorgsky)—Victor 17900-A— A, 28.6; B, 37.1; C, 1.1^5; D, 0.3; E, 1.1; F, 31.8.

APPENDIX C

6 April 1944

The Surgeon General of the Army has authorized a certain group of leaders in the musical world who have organized for the purpose, to make a trial at Walter Reed General Hospital of the possible effect of music in the treatment of certain types of cases.

This musical therapy is not new, having been tried previously in some places, but its application has never been properly or adequately controlled, hence its evaluation as a therapeutic agent has been impossible.

It is the intent of the leaders of the musical therapy group which will operate at Walter Reed General Hospital to attempt to arrive at an unbiased determination of what may be accomplished by properly selected music in various types of cases. The authorities at Walter Reed General Hospital will, in cooperating with the program, so far as possible select a control case comparable in every way to the individual who, in addition to the other hospital procedures, is undergoing the music therapy.

(Signed)   S. U. Marietta
Maj. Gen. U.S.A.M.C.
Commanding General, Walter
Reed General Hospital
Assistant Surgeon General

NOTE: Any statement emanating from Walter Reed General Hospital must be cleared through Major General George F. Lull, of the Surgeon General's Office, before publication.